Communications
in Computer and Information Science 1199

Commenced Publication in 2007
Founding and Former Series Editors:
Phoebe Chen, Alfredo Cuzzocrea, Xiaoyong Du, Orhun Kara, Ting Liu,
Krishna M. Sivalingam, Dominik Ślęzak, Takashi Washio, Xiaokang Yang,
and Junsong Yuan

More information about this series at http://www.springer.com/series/7899

Nina Gunkelmann · Marcus Baum (Eds.)

Simulation Science

Second International Workshop, SimScience 2019
Clausthal-Zellerfeld, May 8–10, 2019
Revised Selected Papers

Springer

Editors
Nina Gunkelmann (iD)
Clausthal University of Technology
Clausthal-Zellerfeld, Germany

Marcus Baum (iD)
University of Göttingen
Göttingen, Niedersachsen, Germany

ISSN 1865-0929 ISSN 1865-0937 (electronic)
Communications in Computer and Information Science
ISBN 978-3-030-45717-4 ISBN 978-3-030-45718-1 (eBook)
https://doi.org/10.1007/978-3-030-45718-1

This Springer imprint is published by the registered company Springer Nature Switzerland AG
The registered company address is: Gewerbestrasse 11, 6330 Cham, Switzerland

Preface

This volume contains the papers selected from the Second Clausthal-Göttingen International Workshop on Simulation Science (SimScience 2019), held during May 8–10, 2019, in Clausthal-Zellerfeld, Germany.

In a world increasingly characterized by complex processes, the simulation of new materials, manufacturing processes, and schedules plays an important role. For this reason, simulation and modeling techniques have become a central activity in many disciplines including natural science, engineering, economics, and social sciences. In the context of the interdisciplinary joint research center Simulation Science Center Clausthal-Göttingen of Clausthal University of Technology and the University of Göttingen, the SimScience 2019 workshop brought together researchers from both industry and academia to present and discuss the latest advances in simulation science. The workshop covered the following topics: simulation and optimization of networks, simulation of materials, and distributed simulation.

From 47 extended abstract submissions, 45 were selected for presentation at the workshop in Clausthal. In a second review round after the workshop, 12 full-length papers of a subset of submissions were accepted for the proceedings. All the selected full-length papers were peer reviewed by two external reviewers based on their qualifications and experience. The selected papers come from researchers from Germany, Indonesia, and the UK.

We wish to thank all co-organizers, the Technical Program Committee members, and all external reviewers for their contributions to SimScience 2019. The submission, registration, and payments were managed by VDE conference services. We also thank Springer for publishing the proceedings of SimScience 2019.

February 2020

Nina Gunkelmann
Marcus Baum

Organization

Joint Organizing Committee

Nina Gunkelmann	TU Clausthal, Germany
Marcus Baum	University of Göttingen, Germany
Gunther Brenner	TU Clausthal, Germany
Jens Grabowski	University of Göttingen, Germany
Thomas Hanschke	TU Clausthal, Germany
Jörg Müller	TU Clausthal, Germany
Anita Schöbel	Fraunhofer ITWM, Germany

Program Committee

Charlotte Becquart	University of Lille, France
Wolfgang Bleck	RWTH Aachen, Germany
Christian Brandl	University of Melbourne, Australia
Valentina Cacchiani	University of Bologna, Italy
Paul Davidsson	University of Malmö, Sweden
Umut Durak	DLR Braunschweig, Germany
Samuel Forest	MINES Paristech, France
Marc Goerigk	University of Siegen, Germany
Patrick Harms	University of Göttingen, Germany
Tom Holvoet	University of Leuven, Belgium
Marco Huber	University of Stuttgart, Germany
Charlotte Kuhn	University of Stuttgart, Germany
Laura De Lorenzis	ETH Zürich, Switzerland
Maximilian Merkert	OVGU Magdeburg, Germany
Kai Nagel	TU Berlin, Germany
Helmut Neukirchen	University of Iceland, Iceland
Bernhard Neumair	KIT, Germany
Narayan Rangaraj	IIT Bombay, India
Ulrich Rieder	Ulm University, Germany
Yudi Rosandi	University of Padjadjaran, Indonesia
Carlos J. Ruestes	National University of Cuyo, Argentina
Siegfried Schmauder	University of Stuttgart, Germany
Rüdiger Schwarze	TU Freiberg, Germany
Thomas Spengler	TU Braunschweig, Germany
Pieter Vansteenwegen	KU Leuven, Belgium
Roberto Gomes de Aguiar Veiga	University of São Paulo, Brazil
Giuseppe Vizzari	University of Milano-Bicocca, Italy
Sigrid Wenzel	University of Kassel, Germany

Peter Wriggers	University of Hanover, Germany
Ronghai Wu	Northwestern Polytechnical University, China
Ramin Yahyapour	GWDG, Germany
Gregory Zacharewicz	IMT Mines Ales, France

Finance Chair

| Alexander Herzog | TU Clausthal, Germany |

Project Support

| Mariel Thieme | TU Clausthal, Germany |

Contents

Optimization and Distributed Simulations

Privacy-Preserving Human-Machine Co-existence on Smart Factory Shop Floors

Alexander Richter[1]([⊠]), Andreas Reinhardt[2], and Delphine Reinhardt[1]

[1] Institute of Computer Science, University of Göttingen, Goldschmidtstr. 7, 37073 Göttingen, Germany
{richter,reinhardt}cs.uni-goettingen.de
[2] Department of Informatics, Technische Universität Clausthal, Julius-Albert-Str. 4, 38768 Clausthal-Zellerfeld, Germany
reinhardt@ieee.org

Abstract. Smart factories are characterized by the presence of both human actors and *Automated Guided Vehicles (AGVs)* for the transport of materials. To avoid collisions between workers and AGVs, the latter must be aware of the workers' location on the shop floor. Wearable devices like smart watches are a viable solution to determine and wirelessly transmit workers' current location. However, when these locations are sent at regular intervals, workers' locations and trajectories can be tracked, thus potentially reducing the acceptance of these devices by workers and staff councils. Deliberately obfuscating location information (*spatial cloaking*) is a widely applied solution to minimize the resulting location privacy implications. However, a number of configuration parameters need to be determined for the safe, yet privacy-preserving, operation of spatial cloaking. We comprehensively analyze the parameter space and derive suitable settings to make smart factories safe and cater to an adequate privacy protection workers.

Keywords: Smart factory · Spatial cloaking · Privacy protection

1 Introduction

The digital revolution has reached industry shop floors around the globe. Besides leading to an optimization of manufacturing processes, it also fundamentally changes the way the employees work. Companies are increasingly relying on the support of industrial robots for assisting in manufacturing processes and goods transport. Autonomous robots have particularly emerged as viable solutions for material transport between storage areas and workplaces. These autonomous robots, also referred to as AGVs, facilitate the autonomous supply of workplaces with materials from warehouses, without the need for human interaction. Sales forecasts for AGVs show an increasing trend for companies to use more AGVs for transport processes in the future [20]. This inevitably leads to an increasing co-existence between humans and robots on the shop floors of smart factories.

© Springer Nature Switzerland AG 2020
N. Gunkelmann and M. Baum (Eds.): SimScience 2019, CCIS 1199, pp. 3–20, 2020.
https://doi.org/10.1007/978-3-030-45718-1_1

The digitalization of manufacturing processes is also changing the way workers work on shop floors. Companies optimize their processes by increasingly promoting the use of wearable computing devices [23] to increase workers' productivity [21,24,27], health [11,12,18], and safety [6]. Wearable devices like smart watches, smart vests, or smart glasses [1,30] already support workers by instructing them or providing them with additional process-related information [21,24]. Moreover, they can contribute to workers' health and safety through their built-in sensors because their data allow for the recognition of user activities, such as walking, standing, sitting, and even the workers' position on the shop floor [2,17].

Connected wearable devices, worn by workers, can inform AGVs about their current locations. This knowledge of the workers' locations prevents AGVs from colliding with humans. There is, however, a downside to a frequent reporting of location information, namely the ensuing threats to the workers' location privacy. Such threats can lead to a reduced acceptance of smart wearables by workers. We therefore investigate the applicability of a location privacy protection techniques in a smart factory scenario. More precisely, we present an extensive simulation study to assess existing trade-offs between AGVs' routing and workers' privacy protection. The rest of the paper is structured as follows. In Sect. 2, we briefly revisit our definition of smart factories and elaborate on the co-existence of workers and AGVs on the shop floors in industrial environments. Section 3 discusses the resulting privacy implications and existing location privacy preserving techniques. We introduce simulation parameters, objectives, and methodology of our study in Sect. 4, before discussing the corresponding results in Sect. 5. At last, Sect. 6 concludes this paper.

2   The Smart Factory

Smart factories are characterized by the presence of AGVs and other robots that contribute to industrial processes [13]. For the AGVs' coordination, a large volume of information is collected and exchanged between participating devices. This enables seamless, safe, and secure interactions between humans, machines, material, and systems [19,22,29]. Human workers still take an important role in smart factories, because of their in-depth understanding of dependencies between process steps and their capability to adequately react to unexpected situations. We thus anticipate that human-machine interactions will continue to exist on shop floors for many years to come.

In this scenario, problems can occur due to the limited space available on the shop floor, though. Often, workers and AGVs need to share the available space (see Fig. 1 for an example). On smart factory shop floors, AGVs are expected to transport materials between machines and workplaces. Their autonomy allows them to collect and deliver items when and where they are needed. Since AGVs move independently between different places, it is of particular importance that AGVs know the positions of the human workers sharing the shop floor, in order to reduce their speed or even completely stop in the case of an impending collision. Diverse options exist to proactively prevent collisions between AGVs and

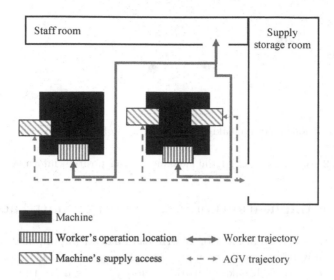

Fig. 1. Sample layout of a shop floor in a smart factory.

workers. Most often, this collision prevention is realized through equipping the autonomous robots with detectors for the human presence, and stopping their operation while a human is present in their immediate environment. Diverse technologies can support human detection. On the one hand, AGVs can be equipped with infrared sensors to detect body heat, radar sensors or laser rangefinders to recognize the shape of human bodies, or cameras to locate humans and anticipate their movements [15]. Particularly, laser rangefinders are often used to detect obstacles on shop floors [10]. However, in that case, the AGVs would not be able to optimize their trajectories in advance. On the other hand, workers can be equipped with wearable devices that periodically broadcast their current position on the shop floor, and thus allow nearby AGVs to stop if they come too close. A strong advantage of the latter type of solutions is their capability to detect workers even when they are not within the camera's field of view. Additionally, such wearable devices can also bring benefits for the workers, such as displaying additional information to accelerate the execution of their tasks [21,24]. The increasing number of smart wearables in companies [23] suggests that companies may want to benefit from the advantages offered by these products in the future. Thus, we follow the latter option, and assume that smart wearables (e.g., smart watches) are worn by the workers in this paper. We further assume that the smart wearables know the workers' location information and can broadcast it wirelessly in order to make it known to the AGVs.

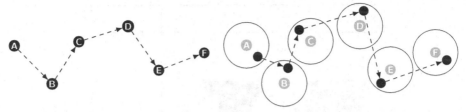

(a) Positions without spatial cloaking.   (b) Positions obfuscated by spatial cloaking.

Fig. 2. Application of spatial cloaking to user position information.

3   Privacy Implications of Wearables in Smart Factories

The regular transmission of location information has strong implications on user privacy. Transmitted information enables the employer to closely monitor workers' routines (e.g., their breaks or work efficiency), categorize them, and eventually even draw inferences about a worker's performance. Even if no unique user identifiers are transmitted, the AGVs receiving the workers' positions and movements can be tracked based on the constant stream of location information. If the AGVs collude with each other by exchanging their own positions and the times when encountering human workers, they can potentially infer users' movements, routes, and identities. This cannot only reduce the acceptance of such solution, but also impacts their compliance with privacy legislation. Consequently, suitable solutions to protect users' privacy must be implemented.

This can be accomplished with a range of different location privacy-preserving techniques. One option is to report multiple "false locations" in addition to the user's actual position [7,8,16]. Thus, the user's location is hidden within the group of fake locations. However, this approach leads to a highly inefficient operation of AGVs because they cannot move within any of the reported areas. Another option is to apply "spatial cloaking" [14], i.e., the intentional reporting of inaccurate data, which we adopt in this work and evaluate its impact when applied in a smart factory setting. Spatial cloaking works as follows: A user's precise location is replaced by a representation of coarser spatial resolution. By way of example, let us look at the diagrams in Fig. 2. When users are required to report their exact positions in regular intervals (as shown in Fig. 2a), their trajectories can be easily traced. In contrast, when spatial cloaking is applied, falsified location points within a definable radius around the users' actual locations are being reported. This is visualized by means of the black markers in Fig. 2b. While these intentional deviations reduce the resolution at which a person can be tracked, they still appear as valid locations and often correctly describe a valid worker trajectory.

Spatial cloaking relies on two key parameters to determine the efficacy of its privacy protection: The radius of the reported area (depicted as a circle in Fig. 2) and the frequency at which reports are being sent. Frequent reporting rates and small reported radii lead to an accurate tracking of human workers, such that the likeliness of collisions with AGVs is greatly reduced. However, the attained degree of privacy protection is similarly low. Conversely, both larger reported radii and a reduced transmission frequency can be used to reduce the precision of the transmitted location information. The latter aspect, i.e., sending reports less frequently, also preserves the energy budgets of the smart wearables better.

However, a change of the reporting transmission frequency has a direct impact on workers' safety, as their actual positions are randomly distributed inside the reported area and unrelated to their heading direction. The risk ensues that workers leave the reported area in-between two successive transmissions, as the reported workers' location information remains the same until the next transmission. Thus, their protection against colliding with AGVs is no longer guaranteed. Figure 3 illustrates three typical cases of reported locations (visualized by the light gray markers) with a radius of three. The worker's actual location is represented by the black markers, while the triangles illustrate the number of steps a worker can take without leaving the reported area and until the location have to be updated to ensure worker's safety. If we assume that the worker's actual location is close to the center of the reported area, the worker can take three steps in the given direction before he/she is leaving the reported area (see Fig. 3a). As the worker is unprotected after leaving the reported area, the location should be updated to ensure the worker's safety. If this is not the case, the worker is considered as unprotected until the next location transmission. In comparison, Fig. 3b illustrates the case when the workers' actual location is close to the perimeter of the reported area and the worker moves inwards. Here, the worker is protected for five steps before the worker leaves the reported area. Within the same transmission frequency as before, the worker is protected for a longer duration (i.e., more steps) in this case due to the actual location and the direction in which he/she moves. In contrast, the worker's actual location is also at an area border in Fig. 3c, but the worker is moving outwards. The worker can only take one step within the same transmission frequency, in which he/she is protected by the reported area. Thus, changing the transmission frequency has an impact on the workers' safety depending on the worker's actual location and the direction he/she wants to move.

4 Simulation Settings

The efficacy of spatial cloaking relies on the choice of its parameters. In order to assist in the choice of these parameters for the safe and efficient operation of workers and AGVs, we conduct an in-depth analysis of the parameter space. More specifically, we analyze how the following factors affect the accuracy of detection workers on a shop floor, and also assess the extent to which workers' privacy is preserved:

1. The maximum allowed deviation between actual and reported location, i.e., the *spatial cloaking radius*.
2. The location information transmission rate, i.e., the *spatial cloaking reporting frequency*.

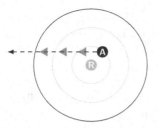  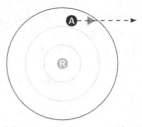

(a) Location reported close to the center of the area.

(b) Location reported at area border, moving inwards.

(c) Location reported at area border, moving outwards.

Fig. 3. Typical cases of location reporting for a cloaking radius of three; A is the worker's actual location and R the reported location. Triangles show the number of steps the worker can take without leaving the reported area.

3. The *transmission success rate* to take into account potential communication loss due to, e.g., channel contention or packet collisions.

For our analysis, we adopt the simulation environment described in Sect. 4.1. We further assume the behavior for both workers and AGVs as detailed in Sects. 4.2 and 4.3, respectively. All simulation results show the average values of three runs with different random seeds.

4.1 Simulation Environment

To evaluate the different parameters, we have created a virtual simulation environment in NetLogo, an agent-based-social framework [25, 28]. In the simulation environment, our smart factory has a size of 128×64 m (a square meter corresponds to a patch, which is the surface unit in NetLogo). Since factories are usually unique in size and organization [3, 9], we have chosen this particular setting to be able to get the first insights. An analysis of the impact of the factory organization on the results is foreseen in future work. In this factory illustrated in Fig. 4, both workers and AGVs move between different areas. While workers visit both their workplaces and the staff room, AGVs roam between workplace storage units and the main storage room. This setting is applied to simulate a manufacturing setting, in which AGVs regularly deliver new materials to the workplaces and move completed items to the main storage. We set the number of workplaces to ten. By doing so, our simulated smart factory can accommodate multiple workers and AGVs with a meaningful degree of activity. During the initialization phase, workplaces are configured to have a pre-defined minimum distance to each other, such that both workers and AGVs are able to move between them easily. Additionally, this means that the trajectories of both AGVs and workers cross in different areas of the smart factory. We set the number of workers to 20 and the number of AGVs to nine. This ensures that each workplace storage is regularly visited by an AGV, and thus the AGVs are almost

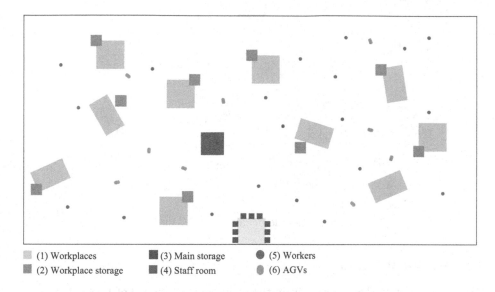

▨ (1) Workplaces	■ (3) Main storage	● (5) Workers
■ (2) Workplace storage	■ (4) Staff room	◉ (6) AGVs

Fig. 4. Sample arrangement of the smart factory used in our evaluation.

constantly in motion on the shop floor. Our simulations terminate when one of the following events occur: (1) when all workers had a near miss with AGVs, which arise when an AGV's and a worker's trajectories cross on the same patch at the same time or (2) when reaching the maximal duration of 36000 s. All simulation settings are summarized in Table 1.

For each random seed, the workplaces, workplaces storages, staff room, AGVs, and workers are distributed randomly inside the factory environment. Only the main storage has a fixed location at the center of all simulated scenarios, as visible in Fig. 4. In order to fully explore the parameter space, we consider transmission success rates between 10% and 100% in increments of 10%. Note that we have chosen this range of transmission rates to cover worst case scenarios, in which, e.g., workers' smart watches may be ill-functioning, and measure their impact on the workers' safety. We, however, expect a normal transmission success rate to be about 90%. Likewise, we vary the size of the reported location between 0 (i.e., no spatial cloaking) and 15 m around a worker. A further enlargement of the radius would only lead to longer AGV waiting times and thus to a significant reduction in productivity. We also vary the workers' location reporting frequency between 1 s and 20 s in increments of 1 s each. A further reduction of the frequency would only lead to an even shorter simulation duration and thus to lower workers' safety due to fewer location updates as explained in Sect. 3. This corresponds to a 10 h working day and thus approximately 2 h over the average working hours inside industrial environments. It hence provides a better comparison with real production environments [5,26]. The previously introduced parameters are chosen to provide a good balance between the workers' and AGVs' motions.

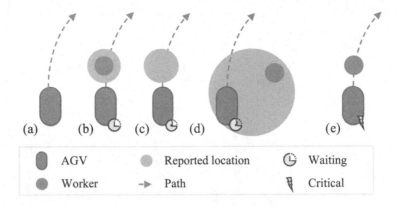

Fig. 5. Possible behaviors of AGVs.

4.2   Worker Behavior

We assume that workers move within the shop floor at a speed of $1.38 \frac{m}{s}$ between the staff room and the workplaces, as the average speed of a pedestrian is approximately $5 \frac{km}{h}$ [4]. The length of stay of the workers at the workplaces and in the staff room is $1 s$. This means that the workers are constantly in motion. At the moment a worker reaches the staff room, he/she selects a random free workplace to visit next. Each workplace can accommodate two workers. In this way it can be ensured that all workplaces in the shop floor are served. Thus, we can evaluate whether the workers' safety can be ensured despite the use of spatial cloaking. If an AGV's and a worker's trajectories cross on the same patch at the same time, the worker is considered to be in shock after this near miss with an AGV, so that he/she cannot work anymore until the end of the shift and is therefore not considered in the simulation scenario anymore. Each worker transmits his/her cloaked location via a smart wearable. This reported location depends on the spatial cloaking radius, the spatial cloaking frequency, and the transmission success rate, which are defined in Sect. 4.1.

4.3   AGV Behavior

We assume that AGVs move within the shop floor at a speed of $2.22 \frac{m}{s}$ meters second between the main storages and the workplace storages. AGVs stay at their destinations for $1 s$ before continuing their journeys, so as to be constantly in motion. Figure 5 illustrates the different AGVs' behaviors on the shop floor. An AGV follows its regular trajectory in absence of any workers' reported locations as shown in the first case, noted (a) in Fig. 5. An AGV immediately stops if it is about to enter a workers' reported location as depicted in cases (b) and (c). In (b), the worker is still located in his/her reported location, while the worker already left the reported location in (c). The latter case can happen when the workers reduce their spatial cloaking reporting frequency. Otherwise, it is also possible that an AGV immediately stops, even if the worker does not cross their

Table 1. Summary of the used simulation parameters.

Parameter	Value
Dimensions of the simulated scenario	128×64 patches
Number of workers	20
Number of AGVs	9
Number of workstations	10
Number of workstation storages	9
Worker velocity	$1.38 \frac{m}{s}$
AGV velocity	$2.22 \frac{m}{s}$
Maximum simulation duration	$36000\,s$
Transmission success probability	$[0.1, \ldots, 1.0]$
Spatial cloaking radius	$[0, \ldots, 15]$ meters
Location transmission frequency	every $[1, \ldots, 20]^{th}$ second
Equivalent real-world distance per patch	1 m
Equivalent real-world time per interval	1 s

path as shown in case (d). When an AGV stops when entering a workers' reported location, it needs to wait until it is able to continue its trajectory. As a result, its productivity decreases. In the absence of workers' reported locations, the AGV continues its work even if a worker crosses its trajectory as seen in case (e). In this case, the AGV performs an emergency braking. We consider that this avoided collision may still fright the worker and impact his/her capacity to work. We hence consider that he/she is unable to work for the remaining of the shift. As a result, this worker is not considered in this simulation run anymore. Please note that this assumption is adopted to allow for an evaluation of the different parameters, and serves as the termination criterion for the evaluation.

5   Simulation Results

We explore the sensitivity of spatial cloaking to the selected simulation settings summarized in Table 1. Across all evaluations, we use both the workers' safety and the AGVs' productivity as metrics. Since the simulation stops as soon as all workers have crossed the path of an AGV, we consider the simulation time as an indicator of the reached workers' safety: The longer the total simulation time, the better for the overall workers' safety. Likewise, the shorter the aggregated time during which AGVs are stopped, the higher the productivity.

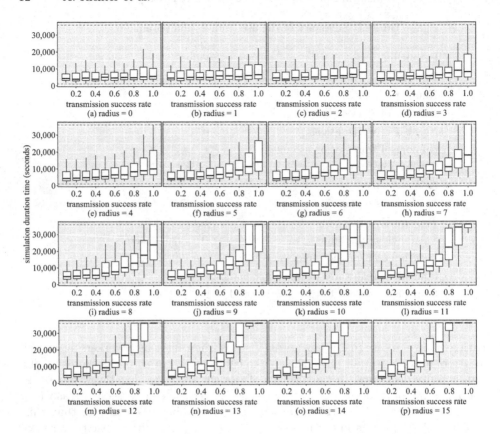

Fig. 6. Influence of spatial cloaking radius on the workers' safety.

5.1 Impact of the Spatial Cloaking Radius

In our first evaluation, we investigate the influence of the spatial cloaking radius on both workers' safety and the AGVs' productivity. Intuitively, a larger radius leads to an increased location privacy protection for the workers and fewer near misses can happen between the AGVs and workers. However, this may decrease the AGVs' productivity.

Figure 6 illustrates the results obtained when varying the spatial cloaking radius from 0 to 15 m. A radius setting of 0 corresponds to reporting the exact workers' location, i.e., the baseline performance without spatial cloaking. In contrast, a radius of 15 corresponds to the most inaccurate location data reporting in our evaluation. The simulation duration is depicted along the y-axes of all box plots, with its upper limit of 36000 s, as per Table 1. On the x-axis, different transmission success rate values are plotted. Boxes show upper and lower quartiles as well as the median value obtained for different values of the spatial cloaking radii. As expected, we observe that greater radii lead to better worker protection, which expresses itself through longer durations of the simulated

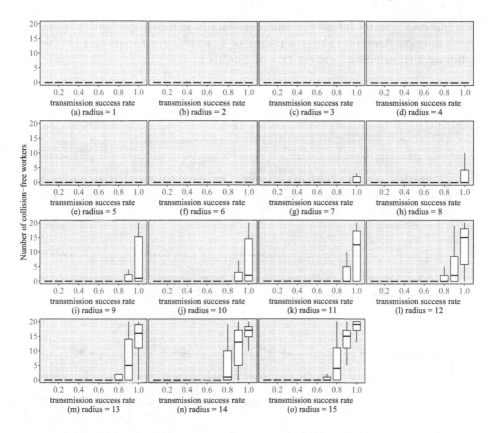

Fig. 7. Influence of spatial cloaking radius on the number of collision-free workers.

settings, because each location report will stop all AGVs within the radius and near misses will thus be are avoided.

A closer look reveals that the results for small radii show almost constant median values for the simulation duration regardless of the transmission success rate. This indicates that spatial cloaking with small radii leads to many near-misses between workers and AGVs. In contrast, increasing the spatial cloaking radius improves the workers' safety. From a radius of 9 m, the median simulation duration is 36000 s, i.e., the full simulation duration. As expected, the simulations confirm that transmission success rate has an impact on the workers' safety. For the same radius of 9 m, packet losses of just 10% lead to a 17% decrease of the simulation duration. Moreover, the medians reveal that the last four radii in the highest transmission success rate reached the full simulation duration. Likewise, the last two radii reached the full simulation duration, also with a 90% transmission success rate. It becomes apparent that the simulation results are sensitive to the transmission success rate. Four more workers remain active (i.e., have not experience near misses with an AGV) when a transmission success rate of 100% is assumed instead of 90%, for a radius of 15 m. When using smaller radii, an even greater number of workers remains collision-free until the end of the simulation duration (11 for a radius of 13 m, 8 for a radius of 14 m) when assuming a transmission success rate of 100%(Fig. 7).

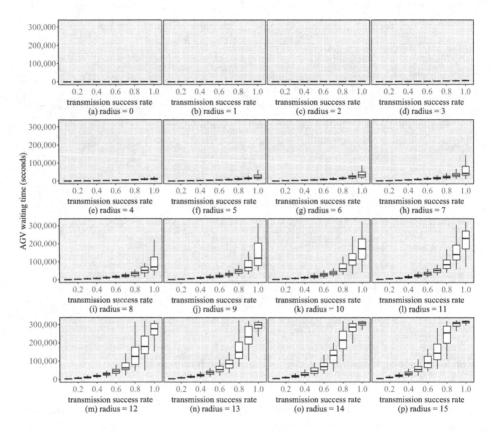

Fig. 8. Influence of the spatial cloaking radius on the AGVs' productivity.

Figure 8 illustrates the influence of the spatial cloaking radius on the AGVs' aggregated waiting time for different transmission success rates. As expected, smaller radii especially for higher transmission success rates lead to higher AGVs' productivity, as the AGVs wait significantly less time. For example, assuming no packet losses, the nine AGVs only need to wait for a total of 3 min for a spatial cloaking radius of 0 m. This time drastically increases to 28 h 21 min when using a radius of 9 m, and 78 h 54 min for 15 m in total. For a very lossy link with an assumed transmission success rate of only 10%, the AGVs' waiting time increase by 61% when the spatial cloaking radius increases from 9 m to 15 m, so that the AGVs seem to stand still almost continuously for larger radii. This confirms the expected trade-off between the size of the spatial cloaking radius for the worker's safety and the AGV's productivity.

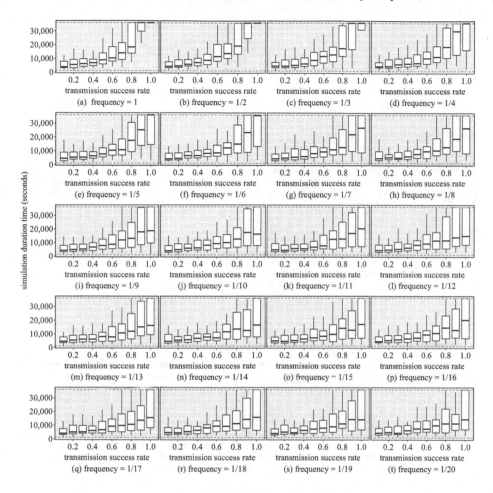

Fig. 9. Influence of spatial cloaking frequency on the workers' safety.

5.2 Impact of the Spatial Cloaking Reporting Frequency

We further analyze the effect of the spatial cloaking reporting frequency on both workers' safety and AGVs' productivity. We assume that a more frequent reporting would lead to an increase of the AGVs' productivity. Moreover, it should lead to fewer near misses between AGVs and workers. However, since their cloaked locations are reported more often, it may be easier for an attacker to infer the workers' actual locations from the reported ones. Figure 9 illustrates the obtained results for spatial cloaking frequency ranging from 1 to 20 s. As expected, we observe that higher frequencies lead to a better workplace safety, as expressed through a longer simulation duration and thus to increased workers' safety. In fact, frequent location updates allow the AGVs to avoid collisions. The graphs indicate that the median values in lower transmission success rates are almost constant. In comparison, for higher transmission success rates and

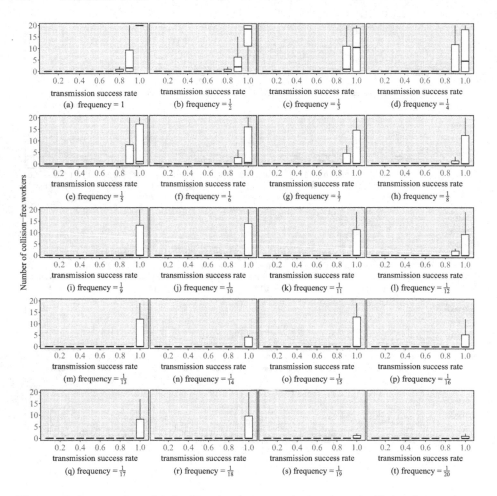

Fig. 10. Influence of spatial cloaking frequency on the number of collision-free workers.

higher reporting frequencies, the simulation duration tend to reach the full simulation duration time of 36000 s. This indicates reporting that spatial cloaking with longer frequencies leads in many cases to unprotected workers, and hence more near misses occur, as the workers take out of their reported locations and are therefore unprotected, until their next location update. From a frequency of 6 s, the median simulation duration achieved is equal to the maximum simulation duration on lossless wireless channels. However, the next lower frequency of 7 s in-between transmissions leads to a reduction of the median simulation duration time by 34.16%, and thus to less workers' safety. Moreover, the results of lower frequencies, especially in conjunction with high transmission success rates, indicate a greater variance. The reason for this is due to the effects of the radius settings, as larger radii increase the workers' safety, especially when using lower reporting frequencies. Workers in small radii leave their protection

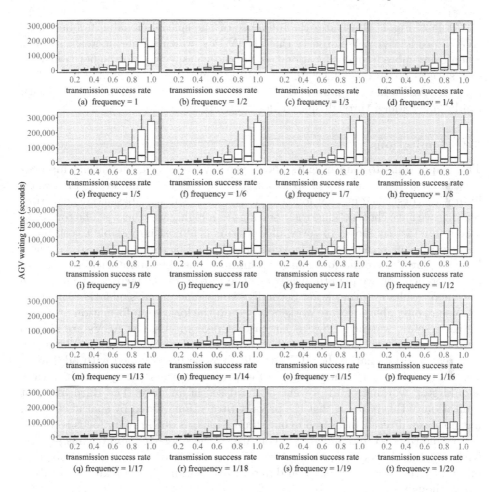

Fig. 11. Influence of spatial cloaking frequency on the AGVs' productivity.

zones significantly faster, which means that they are unprotected for a certain period of time, as mentioned in Sect. 3. For example, at a frequency of 5 s with a radius of 15 m, no worker experiences a near miss with an AGV. However, at the same frequency, but with a reduced radius to 10 m, just 10 workers reached the full simulation duration. While with a further reduction, to a radius of 5 m, no worker is collision-free anymore. In addition, Fig. 10 demonstrates the influence of spatial cloaking frequency on the number of collision-free workers for different transmission success rates. In particular, the change in-between a frequency of 1 and 3 s without any transmission losses lead to 9.5 additional near misses between workers and AGVs in average. Remarkably, even a 10% signal loss lead also in the transmission frequency of 1 s immediately to worse workers' safety as only 1.5 workers achieved the full simulation duration time.

At last, Fig. 11 illustrates the influence of the spatial cloaking frequency on AGVs' productivity. We observe that the AGVs' waiting times increase slightly when using more frequent location transmissions, especially for higher transmission success rates. The longer AGVs' waiting times can be attributed to the fact that the AGVs have to stop more often when workers' location updates (with randomness added through spatial cloaking) are received more frequently.

5.3   Limitations

This paper aims to gain insights about privacy-preserving human-machine co-existence on smart factory shop floors. Our results have, however, three main limitations described in what follows. Firstly, our results are based on the factory settings we have chosen. Since these settings can highly vary between factories [3], a common factory model to cover all scenarios is almost impossible to establish. As a result, our results cannot be generalized to all factories at this stage. They, however, lay the ground for gaining preliminary insights about the feasibility of our approach and we plan to investigate the impact of the factory layout on the obtained results in future work. Secondly, we assume in our simulations that the AGVs are only able to locate workers based on the location information provided by their smart watches. However, in case of near misses, the AGVs are able to perform emergency brakings, so that workers will not be injured. Therefore, in a real-world scenario, the AGVs would likewise be equipped with additional sensors to cater for redundancy and thus, ensure workers' safety if their devices should stop working or to prevent near misses in advance. The third and last limitation of this paper is that the privacy protection offered by spatial cloaking has not been directly measured in our simulations, as our primary focus is to first determine whether and under what conditions the suggested approach is feasible with regards to workers' safety. A detailed analysis of the privacy protection offered by this approach based on different attacker models is planned in future work, though.

6   Conclusion and Outlook

In this paper, we have considered the co-existence of workers and AGVs on smart factory shop floors. Through smart wearable devices configured to regularly transmit position announcements of the human workers, collisions between workers and AGVs can be avoided. Transmitting workers' location information, however, threatens their privacy. We have therefore applied a privacy-preserving solution and measured its effects on both the AGVs' productivity and the resulting workers' safety by means of simulations. While our results are restricted to our tested scenario, they confirm our expectations and allow to quantify the effects of the tested parameters, i.e., the reporting radius, its frequency, as well as the transmission success rate. The results show that the larger the cloaking radii are selected, the better it is for the workers' safety. However, larger radii imply a significant reduction in the AGVs' productivity, which may not be compatible with a real-world deployment. Furthermore, to enhance workers' safety in real-world industrial environments, the AGVs could use previous workers' locations until a new one has been transmitted. Therefore, our work lays the foundation for future explorations of different privacy-preserving solutions. Preserving privacy may help companies to convince workers and works councils to use smart wearables to exploit this potential. Nevertheless, we believe that future research on other location privacy techniques and attacker models is required to fully realize privacy-preserving smart factory environments.

References

1. ABI Research: ABI Research Forecasts Enterprise Wearables will Top US$60 Billion in Revenue in 2022 (2017). https://www.abiresearch.com/press/abi-research-forecasts-enterprise-wearables-will/. Accessed 29 June 2019
2. Bao, L., Intille, S.S.: Activity recognition from user-annotated acceleration data. In: Ferscha, A., Mattern, F. (eds.) Pervasive 2004. LNCS, vol. 3001, pp. 1–17. Springer, Heidelberg (2004). https://doi.org/10.1007/978-3-540-24646-6_1
3. Benjaafar, S., Heragu, S.S., Irani, S.A.: Next generation factory layouts: research challenges and recent progress. Interfaces **32**(6), 58–76 (2002)
4. Carey, N.: Establishing pedestrian walking speeds. Technical report, Portland State University (2005)
5. Carley, M.: Working Time Developments – 2008 (2009). https://www.eurofound.europa.eu/publications/report/2009/working-time-developments-2008. Accessed 12 July 2019
6. Choi, B., Hwang, S., Lee, S.H.: What drives construction workers' acceptance of wearable technologies in the workplace? Indoor localization and wearable health devices for occupational safety and health. Automat. Constr. **84**, 31–41 (2017)
7. Chow, C.Y., Mokbel, M.F., Aref, W.G.: Casper*: query processing for location services without compromising privacy. ACM Trans. Database Syst. (TODS) **34**(4), 1–45 (2009)
8. Chow, C.Y., Mokbel, M.F., Liu, X.: Spatial cloaking for anonymous location-based services in mobile peer-to-peer environments. GeoInformatica **15**(2), 351–380 (2011)
9. Drira, A., Pierreval, H., Hajri-Gabouj, S.: Facility layout problems: a survey. Annu. Rev. Control **31**(2), 255–267 (2007)
10. Golnabi, H.: Role of laser sensor systems in automation and flexible manufacturing. Robot. Comput. Integr. Manuf. **19**(1–2), 201–210 (2003)
11. Gorm, N.: Personal health tracking technologies in practice. In: Lee, C.P., Poltrock, S., Barkhuus, L., Borges, M., Kellogg, W. (eds.) Companion of the ACM Conference on Computer Supported Cooperative Work and Social Computing (CSCW), pp. 69–72 (2017)
12. Gorm, N., Shklovski, I.: Sharing steps in the workplace. In: Proceedings of the ACM Conference on Human Factors in Computing Systems (CHI), pp. 4315–4319 (2016)
13. Grau, A., Indri, M., Bello, L.L., Sauter, T.: Industrial robotics in factory automation: from the early stage to the internet of things. In: Proceedings of the 43rd Annual Conference of the IEEE Industrial Electronics Society (IECON), pp. 6159–6164 (2017)

14. Gruteser, M., Grunwald, D.: Anonymous usage of location-based services through spatial and temporal cloaking. In: Proceedings of the 1st International Conference on Mobile Systems, Applications, and Services (MobiSys), pp. 31–42 (2003)
15. Ilas, C.: Electronic sensing technologies for autonomous ground vehicles: a review. In: Proceedings of the 8th International Symposium on Advanced Topics in Electrical Engineering (ATEE), pp. 1–6 (2013)
16. Kido, H., Yanagisawa, Y., Satoh, T.: An anonymous communication technique using dummies for location-based services. In: Proceedings of the 2nd International Conference on Pervasive Services (ICPS), pp. 88–97 (2005)
17. Lee, S.W., Mase, K.: Activity and location recognition using wearable sensors. IEEE Pervasive Comput. **1**(3), 24–32 (2002)
18. Lingg, E., Leone, G., Spaulding, K., B'Far, R.: Cardea: cloud based employee health and wellness integrated wellness application with a wearable device and the HCM data store. In: Proceedings of the 1st IEEE World Forum on Internet of Things (WF-IoT), pp. 265–270 (2014)
19. Lucke, D., Constantinescu, C., Westkämper, E.: Smart factory-a step towards the next generation of manufacturing. In: Mitsuishi, M., Ueda, K., Kimura, F. (eds.) Manufacturing Systems and Technologies for the New Frontier, pp. 115–118. Springer, London (2008). https://doi.org/10.1007/978-1-84800-267-8_23
20. Murphy, A.: AGV Deep Dive: How Amazons 2012 Acquisition Sparked a $10B Market (2017). https://loupventures.com/agv-deep-dive-how-amazons-2012-acquisition-sparked-a-10b-market/. Accessed 29 June 2019
21. Peissner, M., Hipp, C.: Potenziale der Mensch-Technik-Interaktion für die effiziente und vernetzte Produktion von morgen. Fraunhofer-Verlag Stuttgart (2013)
22. Radziwon, A., Bilberg, A., Bogers, M., Madsen, E.S.: The smart factory: exploring adaptive and flexible manufacturing solutions. Procedia Eng. **69**, 1184–1190 (2014)
23. Schellewald, V., Weber, B., Ellegast, R., Friemert, D., Hartmann, U.: Einsatz von Wearables zur Erfassung der körperlichen Aktivität am Arbeitsplatz. DGUV Forum **11**, 36–37 (2016)
24. Stocker, A., Brandl, P., Michalczuk, R., Rosenberger, M.: Mensch-zentrierte IKT-Lösungen in einer Smart Factory. e & i Elektrotechnik und Informationstechnik **131**(7), 207–211 (2014)
25. Tisue, S., Wilensky, U.: NetLogo: a simple environment for modeling complexity. In: Proceedings of the 7th International Conference on Complex Systems (ICCS), pp. 16–21 (2004)
26. U.S. Bureau of Labor Statistics: Average Weekly Hours of All Employees: Manufacturing [AWHAEMAN] (2019). https://fred.stlouisfed.org/series/AWHAEMAN. Accessed 12 July 2019
27. Weston, M.: Wearable surveillance - a step too far? Strateg. HR Rev. **14**(6), 214–219 (2015)
28. Wilensky, U., Hazzard, E., Froemke, R.: GasLab: an extensible modeling toolkit for exploring statistical mechanics. In: Proceedings of the 7th European Logo Conference (EUROLOGO), pp. 1–13 (1999)
29. Yoon, J.S., Shin, S.J., Suh, S.H.: A conceptual framework for the ubiquitous factory. Int. J. Prod. Res. **50**(8), 2174–2189 (2012)
30. Zebra Technologies: Zebra Study Reveals One-Half of Manufacturers Globally to Adopt Wearable Tech by 2022 (2017). https://www.zebra.com/us/en/about-zebra/newsroom/press-releases/2017/zebra-study-reveals-one-half-of-manufacturers-globally-to-adopt-.html. Accessed 29 June 2019

Dynamic Management
of Multi-level-simulation Workflows
in the Cloud

Johannes Erbel[1]([✉]), Stefan Wittek[2], Jens Grabowski[1], and Andreas Rausch[2]

[1] Institute of Computer Science, University of Goettingen, Goettingen, Germany
{johannes.erbel,grabowski}@cs.uni-goettingen.de
[2] Institute of Software and Systems Engineering, Clausthal University of Technology,
Clausthal-Zellerfeld, Germany
{stefan.wittek,andreas.rausch}@tu-clausthal.de

Abstract. Executing dynamic simulations in a distributed environment allows saving resources and time which is a desired goal in research and industry. One example dynamic simulation is the multi-level-simulation. Here, specific parts of the simulation can be inspected on different levels of detail at runtime. To cope with the changing simulation requirements an elastic and scalable infrastructure is required, as well as an approach adjusting the infrastructure to the simulation needs. In this paper, we enhance a former approach coupling workflows with architectural needs to utilize monitored runtime information and support decision making. Moreover, we demonstrate the concept of executing dynamic simulations over a workflow based approach by dynamically choosing the levels of detail within a supply chain multi-level-simulation.

Keywords: Cloud · Model-driven · Workflow · Multi-level-simulation

1 Introduction

The elastic and scalable nature of cloud computing led to multiple services, tools and standards that have been built around it, providing the opportunity to access computing resources as a utility [1]. One service of interest in research and industry is the execution of simulation workflows on top of this scalable infrastructure. In a former approach we directly coupled workflow tasks to their specific infrastructural needs and reflected them within a *runtime model* which is causally connected to the cloud [8]. Compared to typical workflow management systems [5,26], our approach allows to define, reflect and execute simulation workflows dynamically spawning individual infrastructure and application configurations in the cloud. However, design time concepts to describe the utilization of intermediate results are currently missing as information needs to be gathered and reflected within the runtime representation of the workflow. This runtime information is especially important for dynamic simulations such as *multi-level-simulations* [25]. Here, parts of the system can be simulated on a coarse

© Springer Nature Switzerland AG 2020
N. Gunkelmann and M. Baum (Eds.): SimScience 2019, CCIS 1199, pp. 21–38, 2020.
https://doi.org/10.1007/978-3-030-45718-1_2

or detailed level requiring different amounts of computing resources. Runtime information helps to conclude which coarse grained step should be investigated in detail triggering an automatic reconfiguration of the cloud infrastructure.

In this paper, we investigate how the runtime information of a workflow can be gathered and coupled with its execution in order to support runtime decision making and to meet the dynamic requirements of multi-level-simulations. With the presented approach we enable scientists to define when a detailed simulation should be automatically triggered, as well as trigger them manually at runtime. To reach this goal, we extend an already existing approach utilizing runtime workflow models [8], based on the *Open Cloud Computing Interface* (OCCI) cloud standard [19], and introduce decision making by combining it with monitoring capabilities [6]. Finally, we demonstrate the concept of the presented approach by applying it on an example multi-level-simulation in which parts of a supply chain can be simulated in detail.

The remainder of this work is structured as follows. Section 2 covers the utilized cloud standard and its extensions. Section 3 states problems hindering a dynamic management of multi-level-simulations via workflows. Section 4 presents our approach to solve these problems. Section 5 demonstrates the concept of the presented approach by enacting supply chain multi-level-simulation workflow. Section 6 discusses how the identified problems are tackled and reveals its threats to validity. Section 7 distinguishes our approach to related work. Finally, in Sect. 8, a conclusion is provided as well as an outlook into future work.

2   Foundation

Multi-level-simulations co-simulate a coarse model of the whole system with detailed models required for specific simulation questions [25]. The decision of which detailed models to simulate may change at runtime, requiring a dynamic management of simulation and computing resources. As multi-level-simulations consist of different simulation phases to be performed they can be modeled in form of a workflow. Generally, workflows are expressed either as a *Directed Acyclic Graph* (DAG) or as *Directed Cyclic Graphs* (DCG), if loops are supported [4]. While nodes in these graphs represent the tasks to be computed, links describe a control or dataflow between them. In order to utilize the scalable capabilities of cloud computing [14] within in a scientific workflow, we modeled computational tasks with their required architecture by extending the cloud standard OCCI [8]. In the following we introduce OCCI, as well as an OCCI workflow management system.

2.1   The Open Cloud Computing Interface

The *Open Cloud Computing Interface* (OCCI) is a cloud standard developed by the *Open Grid Forum* (OGF) which allows to access cloud resources in a provider independent way [19]. To achieve this OCCI defines an extensible datamodel for cloud resources and backs it up with a standardized and uniform interface to manage them. The datamodel is based on a generic core model which is easy to extend allowing approaches build around OCCI to be reused by each other. The OCCI interface itself resides next to a clouds proprietary interface and forwards requests to it. Thus, approaches based on OCCI can be applied on any cloud service, e.g., the public *Elastic Compute Cloud* (EC2) from the *Amazon Web Services* (AWS) or private Openstack clouds, as long as a corresponding OCCI interface is available. Conforming to OCCI's datamodel a *metamodel* [13], i.e., a formal language, was developed by Merle et al. [16]. Zalila et al. [27] further enhance this metamodel and the applicability of OCCI by providing an ecosystem that allows to design OCCI models and OCCI extensions, as well as generate code skeletons for OCCI interfaces. In the following the core datamodel of OCCI is introduced, followed by formerly created extensions required to model, monitor, and execute workflows in the cloud.

OCCI Core and Metamodel. The core datamodel of OCCI, shown in Fig. 1, builds the basis for the metamodel described in [16,27]. OCCI's core model is separated into a *classification and identification mechanism* and *core base types*. The core base types, `Entity`, `Resource`, and `Link` form the contents of the system to be designed. Each `Entity` is linked to a `Kind` serving as a descriptor of `Attributes` that can be defined and `Actions` that can be performed on it. Moreover, at runtime an `Entity` can be specified in more detail by attaching `Mixins` to it. These introduce additional `Attributes` and `Actions` available for the `Entity`. The core base types itself are classified by `Kinds` similarly named, together forming OCCI's core extension.

In the metamodel of Zalila et al. [27] two major new elements are introduced. The *Extension* element forms an OCCI extension containing elements from the classification and identification mechanisms, serving as a container for the workflow and monitoring extension introduced below. Moreover, the *Configuration* element is defined serving as top level element containing the `Entity` instances together representing a cloud deployment. It should be noted, that the

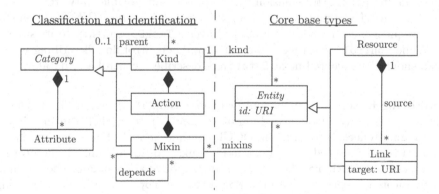

Fig. 1. OCCI core model (adapted from [19]).

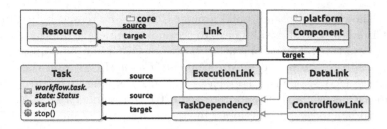

Fig. 2. OCCI workflow extension with Mixins omitted (adapted from [8]).

metamodel introduces multiple new elements helping to formalize OCCI's data-model, e.g., *Finite State Machines* (FSMs) which allow to describe the runtime state of an Entity and how this state is affected by Actions.

While standardized OCCI extensions exist that describe fundamental cloud elements including infrastructure [20], platform [21], and *Service Level Agreement* (SLA) [22] resources, multiple non standardized extensions exists. In former work, we developed a workflow extension [8], as well as a monitoring extension [6], described in the following. These, extensions are required to execute simple workflows and reflect monitoring results within an OCCI runtime model. To model and generate code skeletons from these extensions the OCCIWare [27] toolchain was used, a *Model Driven Engineering* (MDE) framework build around the introduced OCCI metamodel.

Workflow Extension. The workflow extension [8], shown without Mixins in Fig. 2, introduces elements to model a sequence of Tasks in form of a workflow on top of cloud architectures modeled with OCCI. Each Task is a Resource and gives information about its state that is controlled over start and stop actions. When requested, these actions trigger the lifecycle operations of an executable Component to which a Task is linked over an ExecutionLink. Components, defined in the platform extension [21], represent deployable software artifacts placed on OCCI Compute nodes, such as *Virtual Machines* (VMs). To model a task sequence, each Task can be linked over a TaskDependency to its successor. While ControlflowLinks represent a simple control flow describing which Tasks should be executed next, DataLinks describe a control and dataflow.

Monitoring Extension. The monitoring extension [6], shown in Fig. 3, defines elements to specify, deploy, and manage Sensors to gather runtime information and reflect it within an OCCI runtime model. Each Sensor is an Application connected over ComponentLinks to its contained monitoring devices, i.e., DataGatherer, ResultProvider, and DataProcessor. The information monitored is specified within a MonitorableProperty, a Link connecting a Sensor to the Resource it observes. Each MonitorableProperty possesses a monitorable.property and a monitoring.result attribute. These describe the type of information monitored and its result, e.g., the property "CPU" of a VM with the result set to "Critical". Finally, the OCCIResultProvider Mixin can be attached to a `ResultProvider` to define the endpoint of the OCCI interface allowing to reflect gathered monitoring results within the runtime model.

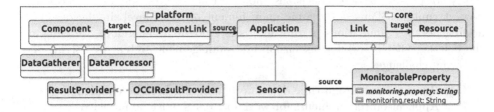

Fig. 3. OCCI monitoring extension (adapted from [6]).

2.2 Runtime Workflow Model Management

In addition to the extensions which allow to design a workflow with OCCI, a `workflow engine`, shown in Fig. 4, is required that performs the OCCI requests in the right sequence enacting the `workflow`. This concept is already sketched in [8] and takes a modeled workflow as input. In general, the engine follows the principles of a self-adaptive *Monitor-Analyze-Plan-Execute-Knowledge* (MAPE-K) control loop [3]. This loop consists of four phases sharing a common knowledge pool to perform adaptive actions [9]. These actions are implemented within the engine's two main components, the `task enactor` and the `architecture scheduler`. Based on the monitored state, reflected within the `runtime model`, and the design time `workflow`, the scheduler analyzes and plans a new required state in form of an OCCI model. The resulting model then only contains the cloud architecture required for the current tasks to be performed. Finally, this model serves as input for a *Models at Runtime* (M@R) engine deriving and executing adaptive OCCI requests bringing the cloud into the desired state [7]. After the adjustment of the cloud deployment, the `task enactor` inspects the `runtime model` and identifies the tasks ready for execution triggering their start action. Within this process the `design` and `runtime model` serve as a knowledge pool for the self-adaptive control loop.

As an example, Fig. 4 shows a workflow with four tasks requiring arbitrary infrastructure deployed one after another. Task A requires one VM and one storage, followed by the tasks B and C together requiring three VMs, which when finished trigger D requiring only one VM.

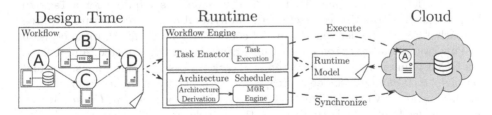

Fig. 4. Runtime workflow modelling (adapted from [8]).

3   Problem Statement

Even though OCCI workflows can be modeled and executed in the cloud, it is currently not possible to define decisions, monitor their results, and adjust the workflow at runtime. Thus, a dynamic management of multi-level-simulations, i.e., adding detailed simulations at runtime, is not possible. Therefore, we enhance the existing OCCI workflow extension and management system by solving the following challenges.

- **C1:** Runtime and design time decision making needs to be coupled.
- **C2:** Application requirements for workflow tasks must be modeled.
- **C3:** Individual cloud architecture models need to be derived at runtime.

4   Dynamic Management of Workflows in the Cloud

In order exploit the full potential of cloud based multi-level-simulation, an integrated concept and adequate tool support is needed that dynamically allocates resources in the cloud according to common workflow concepts. To achieve this, we enhance the former workflow extension [8] and couple it with the capabilities of the runtime monitoring extension [6]. Herewith, we bridge the gap between design time and runtime concepts to execute workflows in the cloud while supporting runtime decision making. In addition to the extension enhancements, we enhance the workflow engine by introducing a *model transformation* [11,15] that generates a model representing the required cloud architecture at each point in time. Combined, we then can decide which part of the workflow should be further processed, e.g., which kind of detailed level simulation should be added, including the provisioning and deployment of the required cloud architecture. In the following the enhanced workflow extensions as well as the adjustments to the workflow engine are discussed.

4.1   Enhanced Workflow Extension

While the former approach allows to define the parallel execution of tasks, a decision of which sub-sequence to execute at runtime is not possible. Therefore, we enhance the introduced workflow extension with new elements and capabilities as highlighted in Fig. 5. While `ExecutionLinks` describe which `Task` executes which `Component`, a description of `Applications` required to run the `Task` could not be modeled. Therefore, we add a `PlatformDependency` link, which allows to define specific `Application` and `Components` required for a `Task` to be performed. Using this connection the individual cloud architecture required for each `Task` can be derived. To model control flow nodes for runtime decision making, we add a `Decision` entity which can be created, adjusted, and deleted via OCCI request. As each `Decision` is a `Task` it possesses the same states and actions, i.e., it can be `started` or `stopped`. The semantic of the decision entity is inspired by already existing languages, e.g., decision nodes within *Unified Modeling Language* (UML) activity diagrams [18] and the *Business Process Model and*

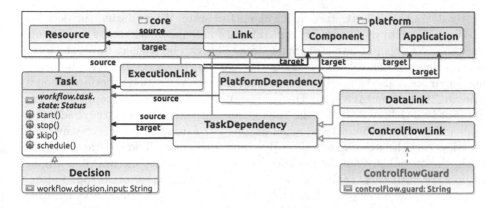

Fig. 5. Enhanced workflow extension for automated decision making.

Notation (BPMN) [17]. In addition to the `Decision` entity, the extension introduces a `ControlflowGuard` Mixin that can be attached to a `ControlflowLink` to define a guard specifying whether a control flow should be followed.

To bridge the gap between design time and runtime decision making, information needs to be gathered and reflected within the runtime workflow model. For this, each `Decision` possesses a `decision.input` attribute to be filled. This is done either manually over an OCCI request or by a monitoring `Sensor` (see Fig. 3) which is connected to a `Decision` over an `ExecutionLink`. This connection ensures that the modeled `Sensor` gets activated for the decision making process. Finally, when the `start` action of the `Decision` is triggered, the `Sensor` is also triggered and its gathered monitoring information is transferred to the `decision.input` and checked against the guards defined in the `ControlflowGuards` of subsequent `ControlflowLinks`.

Each state of a `Task` and `Decision`, and how it can be reached, is defined within a FSM. This FSM is shown as a subset in Fig. 6 and omits the error state, due to clarity, that can be reached by each other state. Together with the `skip` and `schedule` actions, the `skipped` and `scheduled` states are introduced in the enhanced workflow extension. These states define that a `Task` is ready for execution (`scheduled`) or that a `Task` should not be executed even though it is reached within the workflow (`skipped`).

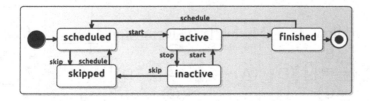

Fig. 6. Finite state machine of a Task with the error state omitted.

By default each `Task` starts in a `scheduled` state. When during the decision making the guard of a `ControlflowLink` does not match the `decision.input`, the target `Task` is put into a `skipped` state. This can be triggered over the `skip` action. Hereby, each subsequent `Task` that can not be reached within the workflow anymore is also skipped. To execute a `Task` a `start` action is performed on it bringing it to the `active` state. During execution a `Task` can be `stopped` making it `inactive`. Otherwise, when the `Task` is done, it traverses to a `finished` state. It should be noted, that a `Task`, when `skipped` or `finished`, can be re-scheduled by triggering the `schedule` action. By combining the individual states of each `Task`, the overall state of the workflow is reflected allowing the workflow engine to operate until each `Task` is either in an error, `finished`, or `skipped` state. Moreover, these allow to analyze and plan the individual architectures required at each point in time, as described in the following.

4.2   Runtime Architecture Derivation

Even though the concept of a self-adaptive workflow engine is already presented, a concept to generate cloud deployments required at runtime is still missing. To tackle this issue, we integrate a model transformation into the architecture scheduling process of the workflow engine. Hereby, the `model transformation`, shown in Fig. 7, generates an OCCI model representing the `required state` in the cloud taking the `runtime model` and `design time model` as input. This grants the transformation access to the current state as well as a representation of the workflow to be executed. Combined, this information is used to derive a model describing the `required state` for the current point in time.

At first, the `model transformation` creates a complete copy of the `design time model` to incorporate the workflow as a high level goal to be reached. In the example shown in Fig. 7 this comprises four tasks and one decision including their architectural needs. Next the transformation checks to which extent this goal has been reached. For this, the transformation updates the state of each task with the information contained within the `runtime model`. Based on the state of the workflow, the transformation identifies each task that is ready to be started, i.e., tasks that are in a scheduled state with its predecessors either skipped or finished. In case of the depicted example, task B is ready for execution as its predecessors are finished, while task C is skipped, and D waits for B to be finished. Finally, the architecture required for this point in time is identified. Within this step each application needed by tasks in a skipped, scheduled, or

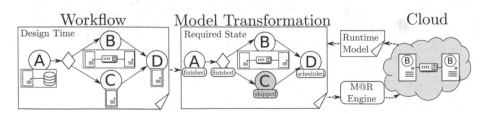

Fig. 7. Model transformation to derive the required cloud state.

finished state are removed from the **required state** model. This includes all connected resources not required by a task ready for execution. To identify these requirements the introduced PlatformDependency is used. Thus, all that is left in the generated model are the elements describing the workflow, as well as the architecture required by each task to be executed. In case of the example, the architecture of task B. Finally, this model serves as input for the M@R engine which derives the required OCCI requests to bring the cloud into the generated **required state**. After the deployment the task enactor identifies that the architectural requirements for task B are met triggering the enactment of the task. It should be noted, that the **model transformation** process is loosely coupled to the rest of the self-adaptive control loop and is thus easily exchangeable.

In the following, we utilize the enhanced workflow extension and engine to execute an example multi-level-simulation scenario in form of an OCCI workflow.

5   Cloud Based Multi-level-simulation

The development of large scale, complex systems requires the application of holistic simulations, as well as highly detailed simulation models to answer specific simulation questions. In multi-level-simulations [25] a coarse level simulation provides the required holistic perspective which is co-simulated with several detailed simulation models. Thus, complexity and computational resources are purposefully and efficiently used, as they are only added when needed resulting in a in dynamic resource requirements. In this section, we demonstrate the concept of executing multi-level simulations through a workflow based approach by applying it on an example scenario. In the following, the multi-level-simulation supply chain scenario is introduced, as well as the workflow model describing it. Thereafter, the execution of the modeled workflow is discussed in detail.

5.1   Example Supply Chain Scenario

In this paper, we developed a multi-level-simulation for a supply chain as a case study. The overall system consists of three sites, as depicted in Fig. 8. The first site is a **warehouse** where the goods stay until a customer orders them. The second site is an **order picking** which combines the goods into shipments. The third site is the **quality assurance** where the picked shipments are subject to an inspection. Here, possible defects are identified before they are finally shipped to the customer. If defective shipments are detected they get send back to the **warehouse** and unpacked again in order to resend them to the **order picking**.

To evaluate the overall supply chain a **coarse simulation** is used. Within this simulation each site is an atomic element not considering the composition of the individual sites. This composition is part of a **detailed simulation**. In case of the supply chain scenario two **detailed simulations** are employed. One simulates the **order picking** and the other the **quality assurance** site. The **order picking** consists of a **buffer A** and a **buffer B** as well as the **picking** station. The **quality assurance** consists of a **sensor**, detecting defects, and a **buffer** to store shipments until they finally depart.

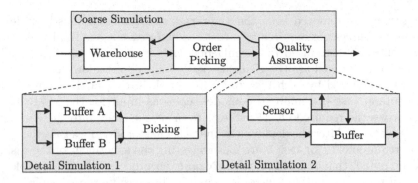

Fig. 8. Multi-level-simulation supply chain scenario.

Each described simulation model is packaged within an executable java archive which can be triggered over a command line interface. While the coarse simulation only requires the amount of simulation steps to be performed, the detailed simulations has to be connected over a remote message interface to the coarse simulation. This requires the individual machines to be connected over a network within the model allowing to automatically determine the address of the machine hosting the coarse simulation at runtime.

5.2 Multi-level-simulation Workflow

To demonstrate our concept, we modeled and executed two OCCI workflows supporting a manual and an automated decision making for the introduced scenario. A conceptual version of this workflow is depicted in Fig. 9 highlighting the task sequence to be executed as well as the provisioned VMs. Within this model, the VMs are connected over a network and host the individual simulations represented by an application consisting of one component. To each component a configuration management script is assigned using the *Model-Driven Configuration Management of Cloud Applications with OCCI* (MoDMaCAO) framework [12] which allows to manage its lifecycle over the OCCI interface. In case of the multi-level-simulation workflow these components represent the individual coarse and detailed simulation applications. Additionally, each workflow task has a component which represents the individual job triggering the execution of the simulation application with specific parameters.

At first, the `coarse` task, simulating the overall supply chain, is executed in isolation for the first 200 time steps with each step being one hour in the simulation. This is done, because it is known that this time is needed for the first orders to arrive and the buffers to be filled. During this part of the simulation workflow, the detailed simulations do not need to be triggered as this setup is of little interest for the performance of the two sites. At this point in time only one VM is required to perform the simulation. After the initial setup interval, one of the two sites, `picking` or `qa`, is selected for further investigation. Hereby, the site to be simulated in detail is chosen based on the results of the `coarse` task.

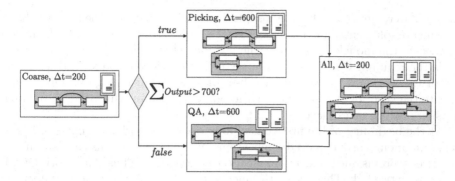

Fig. 9. Supply chain multi-level-simulation workflow.

This can be done either manually or automatically by defining, e.g., a sensor which gathers the results of the **coarse** simulation. Either way, one additional VM is provisioned hosting the application of the detailed simulation.

In case of the presented scenario, we utilized an automated decision making process which investigates the total volume of shipments that left the supply chain. Usually, the detailed simulation of the **qa** would be selected. Only if the **output** volume exceeds $700 \, \mathrm{m}^3$ in shipped packages, the detailed simulation of the **picking** site is selected, because it seems to produce a higher than expected picking overhead. After another 600 time steps, the simulation study is finalized by executing **all** simulations including both detailed simulations for 200 time steps. This approach is efficient in two regards. On the one hand, only the two sections of the system, which are of particular interest, are modeled and executed as a detailed simulation. Moreover, the two detailed simulations are only executed when an in depth simulation is required by the user.

It should be noted, that the dynamic capabilities of the simulation workflow can be further increased by modeling it in more detail, e.g., by adding a loop iterating over the decision making process allowing to continuously decide in short time steps whether a detailed simulation should be added. In the following, we describe how the modeled supply chain multi-level-simulation workflow is executed in detail.

5.3   Executing Multi-level-simulation Workflows

To execute the supply-chain simulation scenario, we enacted the OCCI workflow on a private OpenStack cloud [23]. Hereby, a server application from the OCCIWare toolchain is used that supports the interpretation of OCCI service requests. This server uses **connectors** interpreting incoming OCCI requests, e.g., an infrastructure connector that manages VMs as OCCI Compute nodes. In addition to infrastructure connector for OpenStack, connectors from the MoD-MaCAO framework [12] were used to manage OCCI Application and Components via common configuration management tools. Finally, for this server code skeletons were generated for the workflow extension that got filled with behavior which implements how incoming OCCI service requests should be handled. To execute the modeled workflows, the introduced workflow engine is used that sends the required OCCI requests to server application introduced. Due to the size of the workflow model only a subset of it is shown in Fig. 10. This figure depicts the state of the workflow at the point in time in which the automated decision making takes place.

At the beginning of the first control loop cycle the workflow engine extracts the state of currently deployed applications in the cloud. In this case no infrastructure is provisioned and no application is deployed. Thus, an empty OCCI model is extracted. Thereafter, the model transformation of the architecture scheduling process takes the extracted runtime model and fuses it with the design time workflow. Hereby, all tasks to be executed are part of the generated model. As only the `Coarse` task has no predecessor task, it is marked as ready for execution. Therefore, its required architecture is not deleted from the required runtime state. This comprises the `CoarseVM`, as well as its `Application`, `Component`, and the actual simulation job. Thereafter, the generated model gets deployed via the M@R engine which provisions the `CoarseVM` in the cloud followed by triggering the lifecycle operations of `CApp` deploying the coarse simulation application. Finally, the task enactor identifies that the architectural and workflow requirements to execute the `Coarse` task are met. Thus, the task enactor triggers the `Coarse` task which in turn deploys and starts its executable Component performing the coarse simulation for 200 time steps. While the task is active, the workflow engines control loop waits for it to be `finished`. The executable Component representing the simulation job is not present in the figure below, as it is not deployed anymore at the time of the decision making.

As soon as the `Coarse` task is `finished` the model transformation identifies, that the `Decision` task has to be performed next resulting in a deployment and triggering of its `Sensor` application. As shown in Fig. 10 this `Sensor` consists

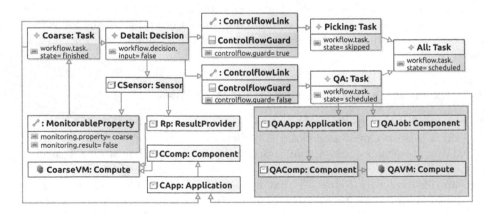

Fig. 10. Subset of the workflow runtime model.

of a `ResultProvider`, with an OCCIResultProvider Mixin, which implements a small monitoring script observing the results of the `Coarse` task by reading the logs of its `Application` and publishes the result to the `MonitorableProperty` within the runtime model. Based on this information, the `decision.input` of the `Decision` is filled and compared against the `controlflow.guard` attributes of connected `ControlflowLinks`.

In this case the `decision.input` is set to `false` resulting in the `QA` task to be executed and the `Picking` task to be `skipped`. After the decision making, the architecture for the still `scheduled` task `QA` is provisioned and deployed. This comprises a new VM on which the detailed simulation is deployed, as well as the simulation job `Component` of the `QA` task, as greyed out in Fig. 10. After the `QA` task has simulated the supply chain for the next 600 time steps, it resides in a finished state. Thus, all workflow requirements to execute the task `All` are met. Within this task each site of the supply chain is simulated in detail adding a new VM to host the picking application and job.

The second modeled workflow, comprising a manual decision making is similar to the one described above. However, instead of a Sensor connected to a Decision, the Decision element waits for the decision.input to be manually filled. For this, a simple service request has to be send to the OCCI interface adjusting the attributes value. It should be noted, due to an MDE approach the created workflow models can be validated. Moreover, instead of an infrastructure connector executing the workflow on a an actual cloud a dummy infrastructure connector can be used to simulate the workflow execution and architecture scheduling without a real execution on a model based level.

6   Discussion

Even though the concept for multi-level-simulation exist, there is no way to dynamically manage them in an automatic and reproducible manner. To solve this issue, we propose the utilization of OCCI workflow models that can be reflected and manipulated at runtime. Within these models, a task sequence can be defined operating on top of a matured and standardized description language for cloud resources. In order to utilize this concept for multi-level-simulations, we identified open challenges, described in Sect. 3, and show how they are solved by modelling and executing an example multi-level-simulation workflow. In the following we discuss how the presented approach solves the identified challenges and reveal its threats to validity.

C1: Runtime and design time decision making needs to be coupled. Even though multiple design time concepts exist that describe the management of control flows within workflows, they do not couple the design time information with a runtime model approach. However, to ensure a casual connection between the runtime model and the executing workflow the information required to perform a decision needs to be gathered and reflected. In this paper, we approached this challenge by coupling the workflow extension with the monitoring extension. Here, we connect a Decision node to a Sensor over an ExecutionLink to serve as its executable. This adds the ability to publish gathered runtime information to the runtime model itself and reflect it within the Decision. Moreover, the enhanced workflow extension introduces control flow guards as well as a skipped state which describes that a task does not have to be executed anymore within an instance of a workflow. With these extension adjustments we enhance the reflective capability of the workflow runtime model and allow it to serve as a knowledge base for the presented workflow engine. Moreover, the causally connection of the runtime model, as well as the utilized standard, allow to easily adjust the input of the Decision node. This way the path to be taken within the workflow can be easily adjusted even at runtime.

C2: Application requirements for workflow tasks must be modeled. In order to enact a task its required infrastructure has to be provisioned and its applications must be deployed. Therefore, a connection to all applications, a task requires to run, must be modeled. We realized this connection within the enhanced workflow extension in form of the PlatformDependency linking a Task to its required Applications. With this link it is known which Applications have to be deployed, as well as the complete infrastructure to run it, as Applications consist of Components deployed on Compute nodes which again may be connected via networks. Especially, this link allows the model transformation within the architecture scheduler to identify which infrastructure is required for each task at each point in time.

C3: Individual cloud architecture models need to be derived at runtime. To solve this challenge, we proposed the utilization of a model transformation within the self-adaptive control loop of the architecture scheduler. This transformation merges the information within the static design time workflow with the runtime model reflecting the actual state of the workflow and the cloud deployment. As a result the transformation generates a new OCCI model describing the required state for each individual step within the workflow. Hereby, the self-reflective capabilities of the states and PlatformDependency introduced in the enhanced workflow extension are used. The generation of the required cloud state in form of a model has the benefit of being self-contained allowing it to be further processed, analyzed, or even simulated before it is passed to the M@R engine adjusting the cloud accordingly. Moreover, model transformations define clear interfaces, as it takes two models as input to form a new one, making it easy to adapt and exchange for other scheduling approaches. With this scheduler we can dynamically provision the individual cloud deployments required throughout the execution of a workflow while not relying on a preconfigured pool of computing resources. However, modeled compute and application resources have to be provisioned and deployed at runtime increasing the time a workflow requires to execute. Still, compared to the execution time of computation tasks these timeframes are negligible and may even be reduced by containerizing the applications required within the workflow.

6.1   Threats to Validity

To manage multi-level-simulations and their dynamic resource requirements the presented approach extends a former approach which couples the modelling of workflows with cloud resources by extending the OCCI cloud standard. Even though this cloud standard grants access to a uniform interface to manage cloud resources, it operates on a low level of abstraction. Therefore, high level concepts can not be defined, e.g., the maximum amount of compute nodes to be deployed during the workflow execution or a dynamic-range of resources available for an individual task. This issue, however, can be solved by coupling the presented approach with a higher abstracted metamodel, e.g., over a model transformation. Still, the standardized interface of OCCI allows to create and maintain a runtime model causally connected to a system granting information about the current state of the cloud deployment. This grants a multitude of benefits, e.g., that the workflow reflection can be used as a knowledge base for self-adaptive control loops. Moreover, as OCCI defines a uniform and standardized interface it is interoperable between different cloud providers and allows already modeled cloud architectures as well as existing tools to be reutilized.

Design time concepts handling control flows within workflows or activities already exist. In this paper we focused on the coupling of this design time elements within a runtime model that reflects a running cloud infrastructure. Although, more design time concept exist, such as loops repeating specific tasks, they are not directly covered within this paper. However, most of these concepts are based on decision making. Thus, the presented approach builds the basis for further features and concepts to be integrated into the runtime reflection of workflows.

Finally, in this paper only an example multi-level-simulation scenario is discussed to demonstrate the dynamic management capabilities of the presented approach. However, the concept can be applied on any kind of computation scenario that can be separated into different tasks to be executed after another. Hereby, the approach allows to model arbitrary architectures for the individual tasks. This in turn, requires knowledge about the required infrastructure itself, as compute nodes need to be specified including the components hosted on them. It should be noted, that the complexity of the modeled cloud architecture depends on the requirements of the individual tasks. This capability describes the major benefit and drawback of the presented approach as users can define arbitrary and shifting infrastructures required for their workflow.

7   Related Work

Multiple concepts and tools exist when it comes to defining and executing workflows, such as Pegasus [5], Taverna [26], or Flowbster [10]. Even though these workflow management systems scale up resources in a predefined cluster, only a few of them consider a connection of individual workflow tasks to arbitrary cloud architectures. In the approach of Qasha et al. [24], e.g., the workflow tasks are coupled to cloud resources described over the *Topology and Orchestration Specification for Cloud Applications* (TOSCA) standard. Still this approach does not follow a model-driven approach. Due to this no runtime model is available reflecting the state of a workflow and the cloud deployment. Beni et al. [2] present a middleware capable of monitoring a workflow at runtime, whereby runtime information is gathered to optimize the infrastructure provisioned for the workflow. However, this information is mainly gathered to improve the workflow for later executions. Their approach is not based on a cloud standard which provide a matured description language for cloud resources, as well as interoperable implementations.

Finally, none of the mentioned approaches provides the capability to reflect the information required for decision making within the runtime model and trigger everything over a uniform and standardized interface. Moreover, we presented an approach and applied it onto a dynamic simulation benefiting from the architecture description and scheduling in terms of the OCCI cloud standard. Thus, already existing OCCI models can be coupled and reused.

8   Conclusion

To dynamically manage multi-level-simulations in the cloud we coupled the simulation with a workflow model based on the OCCI cloud standard. Hereby, we enhanced an already existing approach, which assigns arbitrary infrastructure requirements to workflow tasks, to support runtime decision making. Herewith, we allowed to either automatically or manually choose specific detailed simulations to be added at runtime. The concept to couple runtime and design time decision making is demonstrated within an example supply chain simulation in which detailed simulations of individual packing sites were added at runtime. In the enhanced extension, the information for the decision making is reflected in the runtime model of the workflow. Hereby, we gathered the runtime information over a new link that allows to attach monitoring elements to workflow elements. With this information we decide which workflow path not to follow by transferring them into a skipped state. Finally, to cope with the changing resource requirements of multi-level-simulation applications, we introduced a model transformation that generates a required runtime state of the workflow and cloud by merging the runtime of a workflow with its design time model. This is done every cycle within the planning phase of a self-adaptive control loop and enacted over a M@R engine.

In future work we foster the idea of optimizing parameters for multi-level-simulations during their execution. Hereby, we aim at finding an optimal set of parameters and automatically adjust the required cloud architecture for each part of the simulation. For this, we execute multiple instances of a simulation task covering different parameters. Instead of manually running every simulation separately, we utilize the presented approach to incorporate an optimization heuristic on the design time level of the workflow and couple it with runtime information.

Acknowledgment. We thank the Simulationswissenschaftliches Zentrum Clausthal-Goettingen for financial support.

Availability. The implementation of the presented approach, as well as videos demonstrating the example scenario, can be found at: https://gitlab.gwdg.de/rwm/.

References

1. Armbrust, M., et al.: Above the Clouds: A Berkeley View of Cloud Computing. Electrical Engineering and Computer Sciences, University of California at Berkeley (2009)
2. Beni, E.H., Lagaisse, B., Joosen, W.: Adaptive and reflective middleware for the cloudification of simulation & optimization workflows. In: Proceedings of the 16th Workshop on Adaptive and Reflective Middleware (ARM) (2017)
3. Brun, Y., et al.: Engineering self-adaptive systems through feedback loops. In: Cheng, B.H.C., de Lemos, R., Giese, H., Inverardi, P., Magee, J. (eds.) Software Engineering for Self-Adaptive Systems. LNCS, vol. 5525, pp. 48–70. Springer, Heidelberg (2009). https://doi.org/10.1007/978-3-642-02161-9_3
4. Deelman, E., Gannon, D., Shields, M., Taylor, I.: Workflows and e-science: an overview of workflow system features and capabilities. Futur. Gener. Comput. Syst. **25**, 528–540 (2009)
5. Deelman, E., et al.: Pegasus: a framework for mapping complex scientific workflows onto distributed systems. Sci. Program. J. **13**, 219–237 (2005)
6. Erbel, J., Brand, T., Giese, H., Grabowski, J.: OCCI-compliant, fully causal-connected architecture runtime models supporting sensor management. In: Proceedings of the 14th Symposium on Software Engineering for Adaptive and Self-Managing Systems (SEAMS) (2019)
7. Erbel, J., Korte, F., Grabowski, J.: Comparison and runtime adaptation of cloud application topologies based on OCCI. In: Proceedings of the 8th International Conference on Cloud Computing and Services Science (CLOSER) (2018)
8. Erbel, J., Korte, F., Grabowski, J.: Scheduling architectures for scientific workflows in the cloud. In: Khendek, F., Gotzhein, R. (eds.) SAM 2018. LNCS, vol. 11150, pp. 20–28. Springer, Cham (2018). https://doi.org/10.1007/978-3-030-01042-3_2
9. IBM: An architectural blueprint for autonomic computing. IBM White Paper (2005)
10. Kacsuk, P., Kovács, J., Farkas, Z.: The flowbster cloud-oriented workflow system to process large scientific data sets. J. Grid Comput. **16**, 55–83 (2018). https://doi.org/10.1007/s10723-017-9420-4
11. Kleppe, A.G., Warmer, J.B., Bast, W.: MDA Explained: The Model Driven Architecture: Practice and Promise. Addison-Wesley Professional, Boston (2003)
12. Korte, F., Challita, S., Zalila, F., Merle, P., Grabowski, J.: Model-driven configuration management of cloud applications with OCCI. In: Proceedings of the 8th International Conference on Cloud Computing and Services Science (CLOSER) (2018)
13. Kühne, T.: Matters of (meta-) modeling. Softw. Syst. Model. **5**, 369–385 (2006). https://doi.org/10.1007/s10270-006-0017-9
14. Mell, P., Grance, T.: The NIST Definition of Cloud Computing. National Institute of Standards and Technology (2009)
15. Mens, T., Van Gorp, P.: A taxonomy of model transformation. Electron. Notesin Theor. Comput. Sci. **152**, 125–142 (2006)

16. Merle, P., Barais, O., Parpaillon, J., Plouzeau, N., Tata, S.: A precise metamodel for open cloud computing interface. In: Proceedings of 8th IEEE International Conference on Cloud Computing (CLOUD) (2015)
17. Object Management Group: OMG: Business Process Model and Notation (2011). http://www.omg.org/spec/BPMN/2.0/PDF. Accessed 29 July 2019
18. Object Management Group: Unified Modeling Language (2015). http://www.omg.org/spec/UML/2.5/PDF. Accessed 29 July 2019
19. Open Grid Forum: Open Cloud Computing Interface - Core (2016). https://www.ogf.org/documents/GFD.221.pdf. Accessed 29 July 2019
20. Open Grid Forum: Open Cloud Computing Interface - Infrastructure (2016). https://www.ogf.org/documents/GFD.224.pdf. Accessed 29 July 2019
21. Open Grid Forum: Open Cloud Computing Interface - Platform (2016). https://www.ogf.org/documents/GFD.227.pdf. Accessed 29 July 2019
22. Open Grid Forum: Open Cloud Computing Interface - Service Level Agreements (2016). https://www.ogf.org/documents/GFD.228.pdf. Accessed 29 July 2019
23. OpenStack: Newton (2016). https://releases.openstack.org/newton/. Accessed 29 July 2019
24. Qasha, R., Cala, J., Watson, P.: Dynamic deployment of scientific workflows in the cloud using container virtualization. In: Proceedings of the 8th IEEE International Conference on Cloud Computing Technology and Science (CloudCom) (2016)
25. Wittek, S., Rausch, A.: Learning state mappings in multi-level-simulation. In: Baum, M., Brenner, G., Grabowski, J., Hanschke, T., Hartmann, S., Schöbel, A. (eds.) SimScience 2017. CCIS, vol. 889, pp. 208–218. Springer, Cham (2018). https://doi.org/10.1007/978-3-319-96271-9_13
26. Wolstencroft, K., et al.: The Taverna workflow suite: designing and executing workflows of web services on the desktop, web or in the cloud. Nucleic Acids Res. **41**, W557–W561 (2013)
27. Zalila, F., Challita, S., Merle, P.: A model-driven tool chain for OCCI. In: Proceedings of the 25th International Conference on Cooperative Information Systems (CoopIS) (2017)

On Approximate Bayesian Computation Methods for Multiple Object Tracking

Fabian Sigges[(⊠)] and Marcus Baum

Institute of Computer Science, University of Goettingen, 37077 Goettingen, Germany
{fabian.sigges,marcus.baum}@cs.uni-goettingen.de
http://www.uni-goettingen.de/fusion

Abstract. In this article, we present further results on the use of Approximate Bayesian Computation (ABC) particle filters for multiple object tracking (MOT). Based on our previous work that uses the OSPA distance to select the k-nearest simulated measurements with respect to the actual measurements, we present and evaluate two further ABC variants. The first variant replaces the OSPA distance with a kernel distance, reducing the computational complexity significantly, while the second version exchanges our previous k-nearest-neighbour approach for a distance-based approach, commonly used in ABC algorithms. The algorithms are compared to conventional multiple object tracking algorithms in simulated scenarios with multiple objects.

Keywords: Multi object tracking · Approximate Bayesian Computation · Kernel distance · OSPA · Simulation

1 Introduction

In the last years Advanced Driver Assistance Systems (ADAS) and autonomous driving have gained huge attention from industry, research and society alike. Especially autonomous driving is seen as one of the next big steps in private and public transportation. In order to drive autonomously, a vehicle requires excellent and reliable environmental perception by using e.g. radar, lidar or cameras. One important part of environmental perception is the tracking of detected objects over time, e.g. over multiple frames in a camera video or multiple radar sweeps, which is called multiple object tracking (MOT).

In MOT the problem of measurement association usually arises. For example, if there are two objects and two measurements, it is often unclear which measurement originated from which object. In a Bayesian setting, the unclear measurement association causes the resulting posterior distribution of the object to be multi-modal, one mode for each possible association.

In order to deal with non-linearities, non-Gaussian noise and possibly multi-modal distributions sequential importance resampling (SIR), also called particle filters, have been introduced [18]. Particle filters, as the name suggests, represent the involved distributions by a set of particles, where each particle has an

© Springer Nature Switzerland AG 2020
N. Gunkelmann and M. Baum (Eds.): SimScience 2019, CCIS 1199, pp. 39–51, 2020.
https://doi.org/10.1007/978-3-030-45718-1_3

associated weight. However, MOT using particle filters comes with two major challenges. First, there is the curse of dimensionality, i.e. the required number of particles grows significantly with the number of objects. The second problem concerns the computation of the weights, which requires explicit evaluation of the likelihood function. If the object to measurement association is unclear, the likelihood function involves summing over all possible object to measurement associations. This makes evaluating the likelihood computationally extremely expensive. Both problems have already been tackled before, e.g., by decoupling of states or relaxing the conditions for data association [5].

In other fields, like genetics, biology or psychology, where Bayes' Theorem is used frequently, e.g. for parameter estimation or model fitting, the problem with the likelihood function also exists. For many models the likelihood function cannot even be stated explicitly. To circumvent explicit evaluation of the likelihood, the concept of Approximate Bayesian Computation (ABC) [17] was developed. Here, computation of the likelihood is substituted by simulation. Instead of computing a weight, a measurement is simulated by using a forward or observation model. The simulated measurement is compared to the real measurement using a distance function. Based on the distance, particles are then accepted or rejected.

We initially carried this concept over into the target tracking world in [15], where we gave the general idea of the algorithm and showed some first results, comparing ABC to the global nearest neighbour Kalman filter (GNN-Kf). Additional works using the ABC concept for tracking and navigation, but using different measurement models, are [6], and very recently [10].

In this paper we explore two additional variants of the ABC particle filter and compare them with a particle variant of the Joint Probabilistic Data Association (JPDA) filter [1,7] and a global nearest neighbour Kalman filter (GNN-Kf). While the GNN-Kf serves as a good baseline, a standard JPDA filter serves as an example for an algorithm that uses an expensive approximation of the likelihood function. Especially, we discuss the question if purely simulation based methods can deliver comparable results to likelihood approximation schemes like the JPDA filter.

The paper proceeds as follows. In the next section we give an overview over related work regarding ABC and particle filters. In Sect. 3 the problem formulation is stated and in Sect. 3 the different ABC algorithms are introduced. Sect. 4 is concerned with the evalution of the algorithms and the last section concludes the results.

2   Foundations

This section is concerned with the introduction of the filter variants. The first subsection gives a short review of the development of ABC methods and the second states the problem definition and introduces necessary terms and notation.

2.1 Approximate Bayesian Computation

This subsection is dedicated to a short review of the history of ABC methods. In mathematical biology and related fields like chemistry or psychology many inference problems can be solved by using a Bayesian approach. However, for more complex models the likelihood function cannot even be specified analytically. ABC methods try to brigde this gap via simulation.

The first ABC algorithm that also contains the main idea for the following algorithms is the rejection sampler presented in [12]. Assume we have a measurement z and want to find out the posterior distribution of some parameter x. Then the ABC rejection sampler proceeds as follows

1. Sample a proposal candidate x^* from the prior distribution $p(x)$ of x: $x^* \sim p(x)$
2. Use the likelihood $p(z|x^*)$ to simulate a measurement z^* for the proposal candidate: $z^* \sim p(z|x^*)$
3. Compare the real measurement to the simulated measurement and accept if: $d(z, z^*) < \varepsilon$
4. Repeat this process until there is a sufficient number of accepted particles

Here, $d(\cdot, \cdot)$ is a distance function between two measurements. For a small ϵ the empirical distribution of the x^* is then a good approximation of the real distribution of x. Depending on the choices of, e.g. the ϵ or the generative function, the acceptance rate of the proposal candidates can be very low, up to the point where the application of the algorithm becomes impractical.

The next step in the development was the introduction of ABC with Markov Chain Monte Carlo Methods (MCMC) in [8]. Here, the calculation of the acceptance probability is extended by a check if the simulated measurement is sufficiently close to the real measurement

$$\alpha = \begin{cases} \frac{p(x^*)q(x_i|x^*)}{p(x_i)q(x^*|x_i)} & \text{if } d(z, z^*) \leq \varepsilon \\ 0 & \text{if } d(z, z^*) > \varepsilon. \end{cases}$$

Here, $p(x)$ is again the prior and $q(\cdot|\cdot)$ is the transition kernel to transition from one proposal candidate to the next. With a careful choice of prior and transition kernel, the ABC MCMC methods allows for a much higher acceptance rates than the ABC rejection sampler. However, it can also suffer from the common MCMC issues like a long burn-in phase and chains getting stuck in regions of low probability.

A third option, presented in [16], is the connection of ABC with Sequential Importance Resampling (SIR) samplers. SIR samplers consider a pool of N weighted particles. These are chosen according to their weight and then perturbed by a transition kernel. If the simulated measurement and the real are sufficiently close, the particle is kept, and the new weight is calculated. A good way of choosing the prior and transition kernel is using a number of epsilons of decreasing size $\varepsilon_0 > \varepsilon_1 > \cdots > \varepsilon_m$. For the first iteration the largest ε_0 is used together with the prior. The resulting distribution of particles is then used as the prior for the next iteration with a smaller epsilon. This leads to a smooth transition from the prior to the posterior distribution. As [17] reports the most promising results using SIR filters, we focus on these.

2.2 Problem Formulation

In this subsection, we show how the MOT problem can be stated mathematically and also point out some of the difficulties already mentioned in the introduction. In Fig. 1 three example trajectories are depicted. The objects (green) move from left to right and each time step a measurement (blue) is taken. While the objects are far apart, the association of measurements to objects is clear, however becomes difficult, when they come close together or even cross their paths.

Typically, object tracking is formulated using state-space models. A state-space model usually consists of a process model and a measurement model

$$x_k = f(x_{k-1}, w_{k-1}), \qquad w_{k-1} \sim \mathcal{N}(\mathbf{0}, \mathbf{R}), \qquad (1)$$
$$z_k = h(x_k, v_k), \qquad v_k \sim \mathcal{N}(\mathbf{0}, \mathbf{Q}). \qquad (2)$$

Here, f and h denote the process and measurement model, respectively. The process model is generally used to describe how the state of the object is expected to evolve over time. The measurement model relates the current state of the objects to the measurements. For example, the state of an object might consist of its position and velocity, but the radar measures only distance and angle of the object to the radar. For MOT, the state of the system is $x_k = [x_{k,1}^T, \ldots, x_{k,M}^T]^T$, where x_k denotes the full state of all objects $x_{k,i}$ stacked on top of each other and is denoted in bold, while single object states use standard font. Note that $x_{k,i}$ is usually a vector itself, e.g. position and velocity of the object. The index k denotes the time step and M the number of objects. The vector of measurements is similar to the state vector $z_k = [z_{k,1}^T, \ldots, z_{k,M_c}^T]^T$ a stacked vector of all measurements, where the same notation as for the state applies. Here, M_c already hints that the number of individual measurements and the number of objects might not be equal, due to false detections. As the system is only an approximation to the real dynamics and sensors are never perfect, both models have additional noise w_{k-1} and v_k, which is assumed to be Gaussian distributed with zero mean and covariance $\mathbf{R}$ and $\mathbf{Q}$ respectively. False detections are uniformly distributed over the observed domain, with their number drawn from a Poisson distribution with parameter λ.

In this paper, we are assuming a nearly constant velocity model, see Sect. 4 for more details, for the object dynamics and consider direct measurements of the position of the objects. Under these assumptions the state space model reduces to a linear model with additive Gaussian noise

$$x_k = \mathbf{A}x_{k-1} + w_{k-1}, \qquad w_{k-1} \sim \mathcal{N}(\mathbf{0}, \mathbf{R}), \qquad (3)$$
$$z_k = \mathbf{H}x_k + \mathbf{v}_k, \qquad \mathbf{v}_k \sim \mathcal{N}(\mathbf{0}, \mathbf{Q}). \qquad (4)$$

A major difference between tracking a single object and tracking multiple objects is the data association. If individual measurements are unlabeled, i.e.

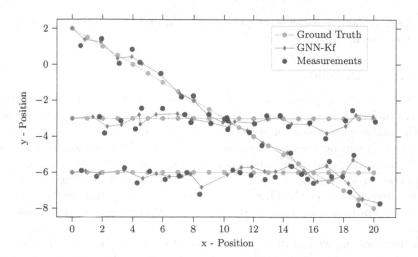

Fig. 1. Example scenario with three objects moving from the left to right and crossing paths, leading to measurement origin uncertainty. In this example, no false detections are shown. (Color figure online)

the measurement to object association is unclear, and every permutation of the individual measurements is possible and assumed to be equally likely, the full stacked measurement model becomes

$$
\underbrace{\begin{bmatrix} z_{k,\pi(1)} \\ \vdots \\ z_{k,\pi(M_c)} \end{bmatrix}}_{=z_k} = \underbrace{\begin{bmatrix} \mathbf{H}_1 & & 0 \\ & \ddots & \\ 0 & & \mathbf{H}_M \\ & \mathbf{0}^{c\times M} & \end{bmatrix}}_{=\mathbf{H}} \cdot \underbrace{\begin{bmatrix} x_{k,1} \\ \vdots \\ x_{k,M} \end{bmatrix}}_{=x_k} + \underbrace{\begin{bmatrix} v_{k,1} \\ \vdots \\ v_{k,M} \\ \tilde{v}_{k,1} \\ \vdots \\ \tilde{v}_{k,c} \end{bmatrix}}_{=v_k}. \tag{5}
$$

Important to note here is the unknown permutation π in the stacked measurement vector z_k. This leads to a likelihood function, which involves summing over all possible associations. Therefore, evaluation of the likelihood function for computing the weights quickly becomes infeasible for a growing number of objects. To complete the model, the $\tilde{v}_{i,c}$ denote the uniformly distributed false detections and we assume a detection probability of one, i.e. the objects are detected in every time step.

3   ABC for MOT

In this section, the use of ABC for MOT is explained. The first subsection explains the basic algorithm and introduces the kernel distance. In the second subsection the kernel distance is replaced with the OSPA distance and the epsilon variant is shown.

3.1   Kernel Distance

In this subsection a variant of the ABC particle filter for MOT is described and one choice of distance function is explained. As we have a fixed pool of N particles and use them as intermediate distribution, but do not calculate weights, the algorithm can be seen as a mixture between the rejection sampler and the SIR. Additionally, instead of using an ε to accept or discard proposal particles, we generate a large pool of N_{prop} particles and then choose the N particles with the lowest distance [4]. These choices were made to guarantee a conceivable and fast computing time, compared to the algorithm presented in the following section.

In the following we will describe the algorithm for one time step to the next. Assume there is a pool of N particles $\{\boldsymbol{x}^i_{k-1}\}^N_{i=1}$ from time step $k-1$. As we are not using weights, particles are drawn uniformly from the set of particles and the process model is applied to get the prediction

$$\boldsymbol{x}^i_k = \mathbf{A}\boldsymbol{x}^p_{k-1} + \boldsymbol{w}^p_{k-1}, \quad \boldsymbol{w}^p_{k-1} \sim \mathcal{N}(\mathbf{0}, \mathbf{R}), \tag{6}$$

$$\text{for } i = 1, \dots, N_{\text{prop}}, \quad p \sim \mathcal{U}(\{1, \dots, N\}). \tag{7}$$

For each of these proposal candidates, a simulated measurement is produced by using the measurement equation from the model

$$\boldsymbol{z}^{i,*}_k = \mathbf{H}\boldsymbol{x}^i_k + \mathbf{v}^i_k, \quad \mathbf{v}^i_k \sim \mathcal{N}(\mathbf{0}, \mathbf{Q}), \quad \text{for } i = 1, \dots, N_{\text{prop}}. \tag{8}$$

Now, a proper distance function is required to compute $d(\boldsymbol{z}_k, \boldsymbol{z}^{i,*}_k)$, the distance between the simulated and the real joint measurement. Since the association between individual measurements and objects is unclear, it is not possible to simply use the Euclidian distance. Instead, a metric has to be used which does not care about the permutation of the individual measurements. One possible option is the kernel distance introduced in [3]

$$d_{\text{Ker}}(\boldsymbol{z}_k, \boldsymbol{z}^{i,*}_k)^2 = \text{const} \cdot \Big(\sum_{l=1}^{M_c} \sum_{j=1}^{M_c} K^{\sqrt{2}b}(z_{k,l} - z_{k,j}) \tag{9}$$

$$- 2 \sum_{l=1}^{M_c} \sum_{j=1}^{M} K^{\sqrt{2}b}(z_{k,l} - z^{i,*}_{k,j}) + \sum_{l=1}^{M} \sum_{j=1}^{M} K^{\sqrt{2}b}(z^{i,*}_{k,l} - z^{i,*}_{k,j}) \Big).$$

Here, K^b is the exponential kernel

$$K^b(x - y) = \exp\Big(-\frac{1}{2} \frac{(x-y)^T(x-y)}{b^2} \Big), \tag{10}$$

where b denotes the kernel width. The kernel distance (9) is derived by taking the kernel distance for distributions and interpreting the multi-object state as a Dirac mixture distribution.

After computation of all the distances between real and simulated measurements, N proposal candidates with the lowest distances are selected as representatives for the state in the current time step.

In terms of computational complexity, the algorithm is very fast, compared to other algorithms for MOT. Computation of the kernel distance requires $\mathcal{O}(M_c^2)$ and as we have to do this for each proposal candidate, we end up with a runtime of $\mathcal{O}(M_c^2 \cdot N_{\text{prop}})$. Commonly MOT algorithms require $\mathcal{O}(M_c^3)$, like the global-nearest neighbour Kalman filter or even exponential runtime in the number of individual measurements as the JPDA filter. However there are approaches, that sacrifice optimality for better runtimes, see e.g., [13] for JPDA approximation schemes, or [9] where softer conditions on data association even lead to a linear-time filter.

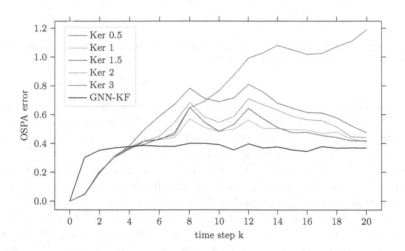

Fig. 2. Comparison for different kernel widths b with low process and measurement noise in the three objects scenario from Fig. 1.

3.2 OSPA Distance

The second ABC particle filter variant was already presented in [15]. There, we introduced the algorithm from the previous section, but instead of the kernel distance, we employed the Optimal Subpattern Assignment (OSPA) distance [14] for the distance calculation. The OSPA distance is a widely used metric for comparing multi-object states and is often used for the evaluation of MOT algorithms. As long as there are no missed detections, a variant of the OSPA distance can be written as

$$d_{\text{OSPA}}(\boldsymbol{z}_k, \boldsymbol{z}_k^{i,*}) = \frac{1}{M} \min_{\pi \in \Pi_{M_c}} \sum_{l=1}^{M} ||z_{k,l} - z_{k,\pi(l)}^{i,*}||_2. \tag{11}$$

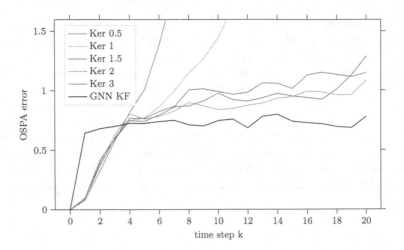

Fig. 3. Comparison for different kernel widths b with high process and measurement noise in the three objects scenario from Fig. 1.

Here, we can see a difference between kernel and OSPA distance in terms of the association. While for the kernel distance every possible association is considered and has an influence on the end result, the OSPA distance matches each object with exactly the one measurement which gives the global minimum.

The algorithm presented in [15] also makes use of the k-nearest neighbour approach. Additionally, for comparison we also implemented an epsilon variant that is more similar to the original ABC algorithm. Here, proposal candidates are generated and their simulated measurements are compared to the real measurement until there are N proposal candidates that fulfill $d_{\text{OSPA}} < \varepsilon$. In terms of complexity, the k-nearest neighbour OSPA variant operates in $\mathcal{O}(M_c^3 \cdot N_{\text{prop}})$, where the cubic term stems from the computation of the minimum in the OSPA using the Hungarian algorithm [11], which has to be done for every proposal candidate. For the epsilon variant, the runtime is the same as for the k-nn variant, however the number of proposal candidates may differ a lot between different scenarios and is highly dependent on the size of ε.

4   Evaluation

In this section, we will evaluate a few properties of the presented algorithms and compare them with a particle variant of the JPDA filter [7] and a global nearest neighbour Kalman filter.

For the evaluation we choose the nearly constant velocity model [2] as the system model, which looks for the two dimensional case as follows

$$\begin{bmatrix} x_k \\ \dot{x}_k \\ y_k \\ \dot{y}_k \end{bmatrix} = \begin{bmatrix} 1 & \Delta T & 0 & 0 \\ 0 & 1 & 0 & 0 \\ 0 & 0 & 1 & \Delta T \\ 0 & 0 & 0 & 1 \end{bmatrix} \begin{bmatrix} x_{k-1} \\ \dot{x}_{k-1} \\ y_{k-1} \\ \dot{y}_{k-1} \end{bmatrix} + \begin{bmatrix} \frac{1}{2}\Delta T^2 w_{k-1,x} \\ \Delta T w_{k-1,x} \\ \frac{1}{2}\Delta T^2 w_{k-1,y} \\ \Delta T w_{k-1,y} \end{bmatrix} \qquad (12)$$

$$w_{k-1,i} \sim \mathcal{N}(\mathbf{0}, \mathbf{R}), \quad i \in \{x, y\}.$$

Here, the state consists of the position and the velocity of the object and ΔT denotes the difference between the time steps $\Delta T = T_k - T_{k-1}$. For the individual measurements we assume that we can directly observe the position of the objects

$$\begin{bmatrix} z_{k,x} \\ z_{k,y} \end{bmatrix} = \begin{bmatrix} 1 & 0 & 0 & 0 \\ 0 & 0 & 1 & 0 \end{bmatrix} \begin{bmatrix} x_k \\ \dot{x}_k \\ y_k \\ \dot{y}_k \end{bmatrix} + \begin{bmatrix} v_{k,x} \\ v_{k,y} \end{bmatrix}, \quad v_{k,i} \sim \mathcal{N}(\mathbf{0}, \mathbf{Q}). \qquad (13)$$

As an error metric we employ the OSPA distance between the mean of the particles $\bar{x}$ and the ground truth data $\hat{x}$

$$E(k) = \left(\frac{1}{M} \min_{\pi \in \Pi_M} \sum_{m=1}^{M} ||\hat{x}_{k,m} - \bar{x}_{k,\pi(m)}||_2^2 \right)^{\frac{1}{2}}. \qquad (14)$$

In Fig. 1 we can see example trajectories of a commonly used scenario. Three objects move from left to right, and one object is crossing path with the other objects, which leads to measurement origin uncertainty. Additionally, the result of a global nearest neighbour Kalman filter is shown in the plot.

As a second comparison algorithm next to the GNN-Kalman filter we use an ensemble Kalman filter variant (ENJPDA) of the Joint Probabilistic Data Association filter [7]. As the ABC algorithms are particle based, we also chose a particle implementation of the JPDA. For data association this filter computes weights to determine how high the influence of each individual measurement on the update should be. Thus, each possible hypothesis has to be taken into account, leading to an exponential runtime in the number of measurements.

The first evaluation concerns the kernel width parameter b. In Figs. 2 and 3 you can see a comparison between different kernel widths in a scenario with three objects and no clutter. For comparison the performance of a GNN Kalman filter is also shown, which is known to provide the optimal solution in linear Gaussian cases, when the associations are known. The figures show that the kernel size should depend on the involved noise, for higher noise a larger kernel size should be taken. In the following simulations we used a kernel size of 1.5 as that seemed to provide the best overall performance in most of the cases.

The next parameter we investigated is the number of required particles N_{pred} and N. In general, the number of required particles highly depends on the scenario and usually more particles result in a better tracking performance, up to a certain point. For the scenario from Fig. 1 we tested how many proposal candidates and how many accepted particles are required. In Fig. 4 you can see the result for the KNN-OSPA variant. As expected, the error decreases slowly for a higher number of particles. For the other variants, the results look similar. In the chosen scenarios the filters do not benefit from more than 50 to 80 accepted particles. In Fig. 5 we can see the results of a comparison of all presented filters in the scenario presented in Fig. 1. Additionally to the measurements a number of false detections is added in each time step. The number of false detections is drawn from a Poisson distribution with parameter λ, where $\lambda = 10$ in this case. The false detections are uniformly distributed over the considered area. For the ABC filters we used $N = 50$ and $N_{\text{pred}} = 200$, while for the ENJPDA we used $N = 100$. All filters are initialized with the correct position and velocity, hence the growing error at the start of the tracking. The error is averaged over 50 runs of the same scenario. The results show that the ENJPDA delivers the best tracking performance, followed by the two OSPA variants, which show a similar performance. The kernel variant also converges to a constant but slightly higher error than the other filters. The increasing error of the GNN-Kf indicates that the filter is not able to handle the false detections and thus diverges from the ground truth trajectories.

The kernel variant is the fastest, followed by the KNN-OSPA variant and the ENJPDA. The epsilon variant is a special case regarding runtime. For decreasing ε the computing time rises very fast, to the point where the algorithm does not finish, because it cannot generate candidate particles within the ε radius. Additional measures, like increasing the system noise after a number of rejected particles might be a solution to this problem.

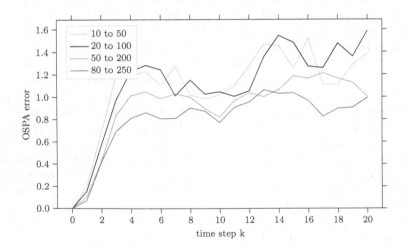

Fig. 4. Comparison for different number of accepted particles N to proposal candidates N_{pred} using the KNN-OSPA variant in the scenario shown in Fig. 1. The other ABC variants show a similar behavior, but are not shown here.

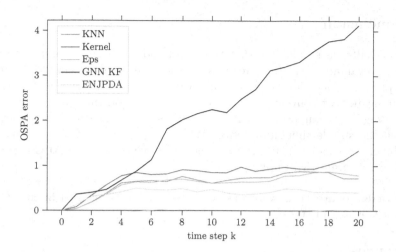

Fig. 5. Comparison of all filters in the scenario shown in Fig. 1, with an average of $\lambda = 10$ uniformly distributed false detections and averaged over 25 simulations.

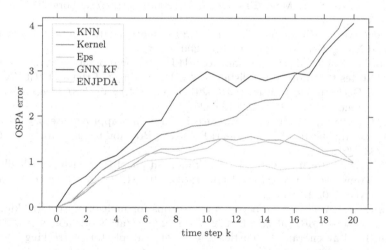

Fig. 6. Comparison of all filters with random trajectories, with an average of $\lambda = 10$ uniformly distributed false detections and averaged over 25 simulations.

In Fig. 6 another comparison between all filters is done. Here, the ground truth trajectories are created using the nearly constant velocity model with a high noise on the acceleration, such that the trajectories almost resemble a random walk. Therefore, the process model used for the prediction step does not coincide with the object motion anymore. The results are very similar to the previous simulation. The ENJPDA works best, followed by the OSPA variants. Both kernel variant and GNN-Kf seem to be losing the tracks quite often, hence the increasing error.

5 Conclusion

A major challenge in MOT is that the likelihood function is computationally complex to evaluate. Therefore, we investigated a class of algorithms that circumvent explicit computation of the likelihood by simulation. However, the experiments indicate that conventional problem-tailored algorithms, like JPDA, have an advantage over our proposed ABC methods. Nevertheless, it is remarkable how well the simple simulation-based ABC methods work. For this reason, it might be interesting in the future to further improve the current ABC approach, e.g. using deterministic sampling.

Acknowledgements. This work was supported by the Simulation Science Center Clausthal/Göttingen (SWZ).

References

1. Bar-Shalom, Y., Daum, F., Huang, J.: The probabilistic data association filter. IEEE Control Syst. Mag. **29**(6), 82–100 (2009). https://doi.org/10.1109/MCS. 2009.934469
2. Bar-Shalom, Y., Kirubarajan, T., Li, X.R.: Estimation with Applications to Tracking and Navigation. Wiley, New York (2002)
3. Baum, M., Ruoff, P., Itte, D., Hanebeck, U.D.: Optimal point estimates for multi-target states based on kernel distances. In: Proceedings of the 51st IEEE Conference on Decision and Control (CDC 2012), Maui, Hawaii, USA, December 2012. https://doi.org/10.1109/CDC.2012.6426189
4. Biau, G., Corou, F., Guyader, A.: New insights into approximate bayesian computation. Annales de l'Institut Henri Poincare, Probabilites et Statistiques **51**(1), 376–403 (2015)
5. Hue, C., Le Cadre, J.P., Perez, P.: Tracking multiple objects with particle filtering. IEEE Trans. Aerosp. Electron. Syst. **38**(3), 791–812 (2002). https://doi.org/10. 1109/TAES.2002.1039400
6. Ickowicz, A.: Approximate bayesian algorithms for multiple target tracking with binary sensors. CoRR abs/1410.4262 (2014). http://arxiv.org/abs/1410.4262
7. Jinan, R., Raveendran, T.: Particle filters for multiple target tracking. Procedia Technol. **24**, 980–987 (2016)
8. Marjoram, P., Molitor, J., Plagnol, V., Tavaré, S.: Markov chain Monte Carlo without likelihoods. Proc. Nat. Acad. Sci. **100**(26), 15324–15328 (2003)
9. Musicki, D., La Scala, B.: Multi-target tracking in clutter without measurement assignment. IEEE Trans. Aerosp. Electron. Syst. **44**(3), 877–896 (2008). https://doi.org/10.1109/TAES.2008.4655350
10. Palmier, C., Dahia, K., Merlinge, N., Del Moral, P., Laneuville, D., Musso, C.: Adaptive approximate Bayesian computational particle filters for underwater terrain aided navigation. In: Proceedings of the 22th International Conference on Information Fusion (FUSION 2019) (2019)
11. Papadimitriou, C.H., Steiglitz, K.: Combinatorial Optimization: Algorithms and Complexity. Dover Publications, New York (1998)
12. Pritchard, J.K., Seielstad, M.T., Perez-Lezaun, A., Feldman, M.W.: Population growth of human Y chromosomes: a study of Y chromosome microsatellites. Mol. Biol. Evol. **16**(12), 1791–1798 (1999)

13. Romeo, K., Crouse, D.F., Bar-Shalom, Y., Willett, P.: The JPDAF in practical systems: approximations. In: Proceedings of SPIE 7698, Signal and Data Processing of Small Targets 2010, vol. 7698, pp. 76981I–76981I–10 (2010). https://doi.org/10.1117/12.862932

14. Schuhmacher, D., Vo, B.T., Vo, B.N.: A consistent metric for performance evaluation of multi-object filters. IEEE Trans. Signal Process. 56(8), 3447–3457 (2008). https://doi.org/10.1109/TSP.2008.920469

15. Sigges, F., Baum, M., Hanebeck, U.D.: A likelihood-free particle filter for multi-object tracking. In: Proceedings of the 20th International Conference on Information Fusion (FUSION 2017), Xi'an, P.R. China, July 2017. https://doi.org/10.23919/ICIF.2017.8009796

16. Sisson, S.A., Fan, Y., Tanaka, M.M.: Sequential Monte Carlo without likelihoods. Proc. Nat. Acad. Sci. 104(6), 1760–1765 (2007)

17. Turner, B.M., Van Zandt, T.: A tutorial on approximate Bayesian computation. J. Math. Psychol. 56(2), 69–85 (2012)

18. Vermaak, J., Godsill, S., Perez, P.: Monte Carlo filtering for multi target tracking and data association. IEEE Trans. Aerosp. Electron. Syst. 41(1), 309–332 (2005). https://doi.org/10.1109/TAES.2005.1413764

Investigating the Role of Pedestrian Groups in Shared Spaces through Simulation Modeling

Suhair Ahmed, Fatema T. Johora$^{(\boxtimes)}$, and Jörg P. Müller

Technische Universität Clausthal, Clausthal-Zellerfeld, Germany
{suhair.ahmed,fatema.tuj.johora,joerg.mueller}@tu-clausthal.de

Abstract. In shared space environments, urban space is shared among different types of road users, who frequently interact with each other to negotiate priority and coordinate their trajectories. Instead of traffic rules, interactions among them are conducted by informal rules like speed limitations and by social protocols e.g., courtesy behavior. Social groups (socially related road users who walk together) are an essential phenomenon in shared spaces and affect the safety and efficiency of such environments. To replicate group phenomena and systematically study their influence in shared spaces; realistic models of social groups and the integration of these models into shared space simulations are required. In this work, we focus on pedestrian groups and adopt an extended version of the social force model in conjunction with a game-theoretic model to simulate their movements. The novelty of our paper is in the modeling of interactions between social groups and vehicles. We validate our model by simulating scenarios involving interaction between social groups and also group-to-vehicle interaction.

Keywords: Pedestrian groups · Mixed traffic · Microscopic simulation

1 Introduction

Over the past years, the shared space design principle has been studied as an alternative to traditional regulated traffic designs. In shared spaces, different types of road users, e.g., pedestrians (often in groups [4]), cyclists and vehicles coexist with little or no explicit traffic regulations. Therefore, road users need to interact with each other more frequently to negotiate their trajectories and avoid conflicts based on social protocols and informal rules. According to [6], we define conflict as "an observable situation in which two or more road users approach each other in time and space to such an extent that there is a risk of collision if their movements remain unchanged."

Studying and understanding heterogeneous road users' movement behaviors under different circumstances and providing realistic simulation models provide

---

S. Ahmed and F. Johora contributed equally to this work.

© Springer Nature Switzerland AG 2020
N. Gunkelmann and M. Baum (Eds.): SimScience 2019, CCIS 1199, pp. 52–69, 2020.
https://doi.org/10.1007/978-3-030-45718-1_4

a good basis for analyzing traffic performance and safety of shared spaces. Modeling mixed traffic interactions is challenging because of the diversity of user types and also of road users behavior. Even in the same user class, each may have a different point of view according to their characteristics.

Pedestrian groups form a large part (70%) of the crowd population [16]. However, previous works on shared space simulation [1,9,22] have been mostly ignoring social group phenomena. In [20], single pedestrian-to-pedestrian group and group-to-group interactions are considered. Works on crowd behavior modeling have paid considerable attention to group dynamics but in homogeneous environments [8,13,23]. In [23], Vizzari et al. proposed a Cellular Automata (CA) based model to describe two types of groups; namely, simple group (a small set of pedestrians) and structured group (a large set of pedestrians which can be structured into smaller subgroups), by considering the goal orientation and cohesion of group. In [8], the classical Social Force Model (SFM) is extended to describe both intra-group and inter-group interactions of pedestrian groups.

Kremyzas et al. also proposed an extension of SFM in [13], called *Social Groups and Navigation (SGN)* to simulate the behavior of small pedestrian groups. In addition to group coherence behavior, SGN can model the situation where a group splits to even smaller groups when needed (e.g. in the overcrowded area) and reforms to re-establish its coherence.

None of these works has considered group-to-vehicle interaction which is challenging because of the heterogeneity in the behavior of different groups. In this paper, we take the first step to address this gap by extending the multiagent-based model described in [9] to incorporate and model pedestrian groups behavior while interacting with vehicles, single pedestrians and other groups.

2 Problem Statements and Requirements

Understanding the movement behaviors of road users in shared spaces is important to model these environments.

To understand the typical interactions of pedestrians and vehicles, in [9], we analyzed the data of a road-like shared space area in Hamburg [20]. From our observation and based on the classification of road user behavior given in [7], we proposed grouping these interactions into two categories, based on the complexity of interaction:

- **Simple Interaction:** Direct mapping of road users perceptions to their actions.
  - **Reactive Interaction:** Road users behave reactively without further thinking to avoid urgent or sudden conflicts. As an example, if a vehicle suddenly stops because a pedestrian suddenly appears in front of the vehicle, then the vehicle behind also needs to perform an emergency break to avoid a serious collision.
  - **Car Following (Vehicle only):** Empirical observation states that although there is no defined lane for vehicles in shared spaces, they follow the vehicle in front as they are driving into an assumed lane [1].

- **Complex Interaction:** Road users choose strategy among different alternative strategies.
  - • **Implicit Interaction:** Road users choose their best action by predicting others' action. As a real-life example, when a pedestrian crosses a road with a high speed, an approaching vehicle might predict that the pedestrian will continue to walk, so the driver will decelerate to avoid collision.
  - • **Explicit Interaction:** Road users also interact with others using hands or eye contact as a means of communication.

In this paper, we analyze the properties of pedestrian groups and their interaction with others. The term 'Group' here does not mean a couple of random people that happen to walk close to each other. According to [16], we define a group term as "it is not only referring to several proximate pedestrians that happen to walk close to each other, but to individuals who have social ties and intentionally walk together, such as friends or family members". We describe pedestrian groups by the following features:

- *Size*: The most frequent group sizes vary between two to four members, whereas groups comprising five or more members are rare according to [16].
- *Goal Position*: Normally all group members walk towards predefined a common destination.
- *Coherence*: Coherence is an important criterion of groups; in other words, group members manage to stay together. However, if the group members split temporarily for any reason, the faster members would wait at a safe point until other members reach that point and the group becomes coherent again [10].
- *Speed*: The average speed of pedestrians in a group is dependent on the group size; bigger groups are slower compared to small groups and the relation between group size and average speed of group is linear [16].
- *Clustering Option*: We observe from real shared space data and also from the literature on group dynamics [5,8] that social groups frequently split if the group size is greater than 3.

In case of group-to-group interaction, groups usually deviate from their current path instead of decelerating or accelerating, to avoid collision. Interactions between a group of pedestrians and a vehicle are observed to happen typically in the following way: group members follow a temporary leader who decides on action (decelerate, accelerate or deviate) first or they split into subgroups (groups split even while walking [8]), subgroups interact with the vehicle individually and re-form afterwards. The leader reacts in a similar way as a single person would, but by taking the group size and positions of the group members into consideration.

In Fig. 1, we depict an imaginary scenario to illustrate group-to-vehicle interaction behavior: where a car Car_1 (moving to the *right* direction) interacts with a pedestrian group $G_1 = \{M_1, M_2, M_3, M_4, M_5\}$ (crossing a road from *bottom to top* direction). Lets say, in this case, the outcome of their interaction is that Car_1

will wait and all or some members of G_1 will continue. If G_1 is a non-clustered group with one leader member who makes decision in any situation and all other members follow his/her decision, then the interaction between G_1 and Car_1 is executed as shown in the following steps:

- 1: Car_1 starts decelerating to let the group cross and G_1 starts crossing the road.
- 2: Car_1 eventually stops and G_1 continues crossing the road.
- 3: G_1 almost crossed the road by maintaining group coherence i.e. all group members move together as a single group without splitting into smaller subgroups. Car_1 starts moving as G_1 is at a safe distance.
- 4: Both Car_1 and G_1 continue moving towards their destinations.

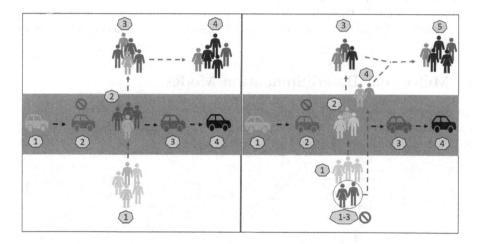

Fig. 1. Non-clustered group (left) vs. clustered group (right) interaction with a vehicle. Here, the numbers (1–5) represent the interaction steps and the lighter to darker shades of colors represent positions of car and pedestrians at different steps (lightest: first step and darkest: last step). The group leader and last member are visualized by blue and orange colors respectively. (Color figure online)

Whereas, if G_1 is a clustered group, then the execution of interaction between G_1 and Car_1 is performed in the following steps:

- 1: Car_1 starts decelerating to let the group pass first. G_1 splits into two subgroups and the subgroup which contains the original group leader, $G_{1.1}$ starts crossing the road. Whereas, the other subgroup $G_{1.2}$ waits (other possible actions for $G_{1.2}$ are continue walking or deviate) for Car_1 to pass.
- 2: Car_1 eventually stops, $G_{1.1}$ continues crossing the road and $G_{1.2}$ waits.
- 3: $G_{1.1}$ crosses the road and waits for $G_{1.2}$ to re-form the group, and Car_1 starts moving as both subgroups are at safe distances.
- 4: Car_1 continues driving, $G_{1.1}$ waits for $G_{1.2}$ and $G_{1.2}$ starts crossing the road.

– 5: $G_{1.1}$ and $G_{1.2}$ meet, re-form the group G_1 to reestablish their group coherence and then G_1 continues moving towards its destination.

We model the intra-group interaction and interaction of pedestrian group with single pedestrian or other groups (assumed to be similar to pedestrian-to-pedestrian interaction) as simple interaction, whereas we model the pedestrian group-to-vehicle interaction both as simple and complex interactions. The **simple interactions** can be modeled using force-based [7] or cellular automata models [17]. However, decision-theoretic models such as multinomial logit model [7] or game theoretic model [3,11,14,21] are more efficient for representing **complex interactions** [7]. In a logit model, a decision maker makes a decision regardless of others' decision based on present data [19], whereas, in a non-cooperative game-theoretic model, each decision maker includes predictions of other players' decisions in its own decisions.

We note that in this paper we only consider single pedestrian, groups of pedestrians and vehicles as road users; modeling *explicit* interaction is a topic of future work.

3   Multiagent-Based Simulation Model

We integrate pedestrian group dynamics to our existing simulation model [9] which comprises of three modules: trajectory planning, force-based modeling, and game-theoretic decision-making as visualized by Fig. 2.

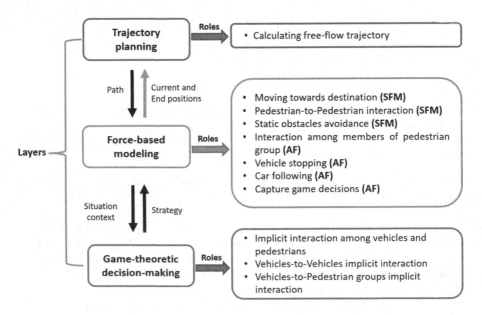

Fig. 2. Conceptual model of road users' motion behaviors. AF means additional force.

3.1 Trajectory Planning Layer

The trajectory planning module computes free-flow trajectories of each road user by considering static obstacles such as the buildings boundaries or trees in the shared space environment. We convert our simulation environment into a visibility graph by connecting outline vertices of the obstacles, to perform the A* trajectory planning algorithm [16]. We only connect two vertices if they are visible to each other and their resulting edge does not collide with any outline of any obstacle [12]. We also add the origin and destination points of each road user to the graph. After planning the trajectories of road users, we adjust the position of their inner path vertices to capture the fact that humans tend to keep some distance from obstacles.

3.2 Force-Based Modeling Layer

We use the social force model (SFM), introduced by Helbing et al. [7], to model simple interactions of road users. In the classical SFM, the movement of a pedestrian is controlled by a set of simple forces, which are represented in Eq. 1. These forces reflect the inner motivation of a pedestrian to perform certain actions based on the circumstances.

$$\frac{d\overrightarrow{v}_a}{dt} := \overrightarrow{f}_\alpha^o + \Sigma \overrightarrow{f}_{\alpha B} + \Sigma \overrightarrow{f}_{\alpha \beta} \tag{1}$$

We apply the classical SFM to capture the driving force of road users towards their destination ($\overrightarrow{f}_\alpha^o$), their repulsive force towards static obstacles ($\overrightarrow{f}_{\alpha B}$) and towards other pedestrians ($\overrightarrow{f}_{\alpha \beta}$). We extend SFM to model car following interaction among vehicles ($\overrightarrow{I}_{following}$) and reactive interaction ($\overrightarrow{I}_{stopping}$) of vehicles towards pedestrians by decelerating to allow pedestrians to pass first. $\overrightarrow{I}_{stopping}$ only occurs if pedestrian(s) or pedestrian group(s) has already started moving in front or nearby to the vehicle. The details of these forces is in [9].

We select the Social Groups and Navigation model (SGN) of [13] and the model of Moussaïd et al. [15] to capture the intra-group behavior of pedestrian groups. Both these models are extensions of SFM [7]. To model vehicle-to-pedestrian groups interaction, we extend these both models, see Subsect. 3.4.

The group force term $\overrightarrow{f}_{group}$ defines the interaction among group members. It combines two forces, namely, visibility force $\overrightarrow{f}_{vis}$ and attraction force $\overrightarrow{f}_{att}$[1], to help group members to stay connected and maintain the group structure. $\overrightarrow{f}_{vis}$ stands for the desire of a pedestrian to keep his/her group members within his/her field of view (FOV, see Fig. 3) and $\overrightarrow{f}_{att}$ attracts any group member (except the leader) to the centroid of the group, when the member exceeds a calculated threshold value d unless the respective member reaches his/her goal and as a result his/her desired velocity $V_{desired}$ becomes zero.

---

[1] vector + vector = vector; point + vector = point.

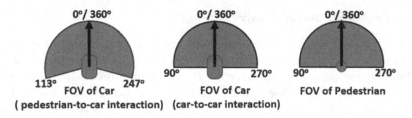

Fig. 3. Road users field of view.

$$\overrightarrow{f}_{group} = \overrightarrow{f}_{vis} + \overrightarrow{f}_{att} \tag{2}$$

$$\overrightarrow{f}_{vis} = S_{vis} * \theta * \overrightarrow{V}_{desired} \tag{3}$$

$$\overrightarrow{f}_{att} = \begin{cases} S_{att} * \overrightarrow{n}(A_{ij}, C_i), & \text{if } dist(A_{ij}, C_i) \geq d \text{ and } \overrightarrow{V}_{desired} \neq 0 \\ 0, & \text{otherwise.} \end{cases} \tag{4}$$

Here, S_{vis} and S_{att} are global strength parameters, θ is the minimum angle between every two members in a pedestrian group G_i that they should maintain to stay within each other's FOV, $\overrightarrow{V}_{desired}$ denotes the desired velocity of any member A_{ij} of G_i, and $\overrightarrow{n}(A_{ij}, C_i)$ represents the distance between A_{ij} and the centroid of the group (C_i), normalized in unit length. The C_i is defined as follows:

$$C_i = \frac{1}{|G_i|} \sum_{1 \leq j \leq |G_i|} x_{ij} \tag{5}$$

Here, $|G_i|$ denotes the total number of members in G_i and x_{ij} depicts the position of any member A_{ij} of G_i.

Execution of the game module decisions ($\overrightarrow{I}_{game}$) are also handled in this module, for example, if the result of a game-theoretic interaction between a vehicle driver and a group of pedestrians is that the groups can continue and the vehicle should wait (see also Subsect. 3.3), the respective actions of all users will be executed in this module. In this paper, we consider the continue, decelerate and deviate (for single or group of pedestrians only) as feasible strategies for road users and model these strategies as follows:

– Continue: Any pedestrian α crosses vehicle β from the point $p_\alpha = x_\beta(t) + S_c * \overrightarrow{e}_\beta$, if $line(x_\alpha(t), E_\alpha)$ intersects $line(x_\beta(t) + S_c * \overrightarrow{e}_\beta, x_\beta(t) - \frac{S_c}{2} * \overrightarrow{e}_\beta)$, otherwise free-flow movement is continued. For vehicles, they continue their free-flow movement without any deviation. Here, scaling factor which depends on vehicle's speed is denoted by S_c, $\overrightarrow{e}$ is the direction vector, $x(t)$ and E represent current and goal positions respectively.

– Decelerate: Road users decelerate and in the end stop (if necessary). For pedestrians, $newSpeed_\alpha = \frac{currentSpeed_\alpha}{2}$ and in case of vehicles, $newSpeed_\beta = currentSpeed_\beta - decelerationRate$.

Here, $decelerationRate = \begin{cases} \frac{currentSpeed_\beta}{2}, & \text{if } distance(\alpha,\beta) \leq D_{min}, \\ \frac{currentSpeed_\beta^2}{distance(\alpha,\beta)-D_{min}}, & \text{otherwise.} \end{cases}$

D_{min} is the critical spatial distance.

– Deviate: A pedestrian α passes a vehicle β from behind from a position $p_\alpha = x_\beta(t) - S_d * \overrightarrow{e}_\beta(t)$ and after that α resumes moving towards its original destination. However, as long as β stays in range of the field of view (FOV) of α, α will keep moving towards p_α. Here, S_d is a scaling factor. The value of S_c, D_{min}, and S_d need to be calibrated.

The overall resulting behaviors of pedestrians and vehicles are presented in Eqs. 6 and 7 respectively.

$$\frac{d\overrightarrow{v}_a}{dt} := \left(\overrightarrow{f}_\alpha^o + \Sigma\overrightarrow{f}_{\alpha B} + \Sigma\overrightarrow{f}_{\alpha\beta} + \overrightarrow{f}_{group}\right) \text{ or } \overrightarrow{I}_{game} \tag{6}$$

$$\frac{d\overrightarrow{v}_a}{dt} := \overrightarrow{f}_\alpha^o \text{ or } \overrightarrow{I}_{following} \text{ or } \overrightarrow{I}_{game} \text{ or } \overrightarrow{I}_{stopping} \tag{7}$$

The game-theoretic decision has priority over the decision of this module for both user types, except for $\overrightarrow{I}_{stopping}$ of vehicles.

3.3   Game-Theoretic Decision Layer

The game-theoretic decision module handles the implicit interactions between two or more road users. We use a sequential leader-follower game called Stackelberg game to handle these interactions, in which the leader player acts first and the followers react based on the leader's action to maximize their utility [21]. We set the number of leaders to one and there can be one or more followers for any individual game. The vehicle (faster agent) is chosen as the leader in case of pedestrian(s)-to-vehicle and vehicle-to-group interactions, whereas in case of pedestrian(s)-to-multiple vehicles and vehicle-to-vehicle interactions, we pick one of these vehicles as the leader randomly. Only one game is played for each implicit interaction and the games are independent of each other.

As projected by Fig. 4, to calculate the payoff matrix of the game, firstly, we ordinary value all actions of the players from −100 to 4 (here, positive values are preferred outcome) with the assumption that reaching their destination safely and quickly is their main preference. Secondly, to capture courtesy behavior and situation dynamics, eleven relevant observable factors are taken into account. Among these factors, two are especially for groups-to-vehicle interaction. These factors are defined in the following as Boolean (1, 0) variables $x_1 \ldots x_{11}$ and used to calculate a set of parameters $F_1 \ldots F_{26}$, which are impacts of these factors. Let α be a road user who interacts with another user β:

Car \ Ped	Continue	Decelerate	Deviate
Continue	-100, -100	$4 + C_c$, P_D	$4 + C_c$, $1 + F_{26} + P_{dev}$
Decelerate	C_D, $4 + P_c$	-50, -50	C_D, $F_{16} + P_{dev}$

(a)

Car1 \ Car2	Continue	Decelerate
Continue	-100, -100	$2 + CC_c$, CC_D
Decelerate	CC_D, $2 + CC_c$	-50, -50

(b)

Pedestrian-to-Car:

$C_c = F_1 + F_4 + F_5 + F_9 + F_{13} + F_{22} + F_{23} + F_{24}$
$C_d = F_2 + F_{10} + F_6 + F_{18} + F_{25}$
$P_c = F_1 + F_7 + F_{19} + F_{14} + F_{26}$ **Car-to-Car:**
$P_d = F_2 + F_8$ $CC_c = F_1 + F_{11} + F_{20}$
$P_{dev} = F_3 + F_{15} + F_{17}$ $CC_d = F_2 + F_{12} + F_{21}$

(c)

Fig. 4. The complete payoff matrices of pedestrian/group-to-vehicle and vehicle-to-vehicle interactions with all considered actions. (a) Pedestrian/group-to-vehicle Interaction (b) Vehicle-to-vehicle Interaction (c) Impacts of situation dynamics

- x_1: has value 1, if current speed of β, $S_{current} < S_{normal}$. This factor determines the value of F_1 to F_3.
- x_2: has value 1 if number of active interactions $< N$. This factor decides the value of F_4.
- x_3: has value 1 if α already stopping to give way to another user β'. This decides the value of F_5 to F_8 and F_{16}.
- x_4: has value 1 if α is a car driver following another car β'. This factor determines the value of F_9 to F_{12}.
- x_5: has value 1 if path deviation does not result long detour for α (pedestrian): $(\theta_{\vec{e}_\beta \hat{n}_{\alpha\beta}} > 58°$ and $\theta_{\vec{e}_\beta \hat{n}_{\alpha\beta}} \leq 113°)$ or $(\theta_{\vec{e}_\beta \hat{n}_{\alpha\beta}} \geq 247°$ and $\theta_{\vec{e}_\beta \hat{n}_{\alpha\beta}} < 302°)$. This factor decides the value of F_{13} to F_{15}.
- x_6: has value 1 if α is a car driver followed by another car β'. This factor determines the value of F_{17}.
- x_7: has value 1 if distance$(\alpha, \beta) < M$, then α (car) is unable to stop. This factor determines the value of F_{18} to F_{19}.
- x_8: has value 1 if α as a car is in a roundabout. This factor determines the value of F_{20} and F_{21}.
- x_9: to consider uncertainty in driving behavior, the value of F_{22} generates randomly.
- x_{10}: has value 1 if the leader α of a group is in waiting state. This factor determines the value of F_{23}.

– x_{11}: has value 1 if a pedestrian α is in a group. This factor determines the value of F_{24}, F_{25} and F_{26}.

The same variable x_i is often used to determine multiple parameters e.g. x_1 calculates the value of F_1, F_2 and F_3, because, the same value of x_i has different influences on different strategies of road users. As an example, if the speed of a pedestrian is less than her average speed, then x_1 has positive influence on strategy decelerate (F_2) and negative influence on continue (F_1). To analyze the behavior of our model, we perform a sensitivity analysis of the parameters F_1 to F_{26}, N and M on a certain amount of interaction scenarios. However, as we have not calibrate these parameters with a significant amount of real interaction scenarios, we do not present these parameters values in this paper. As part of our future work, we will focus on automated calibration and validation of these parameters. More details regarding these modules, game solving, interaction categorization and modeling can be found in [9].

3.4   Interaction Handling of Pedestrian Groups

We have assigned some properties for each group which are essential to define, model and simulate the group behaviors: a specific ID, a leader L_{ij}, a boundary member Lm_{ij} (the one with the largest distance to the leader) which is changeable during simulation [13] and group splitting behavior; if the size S of any group G is as such that S $\geq$ 3, then G splits into subgroups ($G_1, .., G_n$) based on the probability P defined as $P_{base} + (S - 3) * \alpha$. The calibration of the values of P_{base} and α is part of our future research.

In our model, we choose L_{ij} in three different methods: namely, the nearest member to the competitive vehicle, to the group destination or to the road boarders. A group is called a coherent group if the distance between the last member Lm_{ij} and leader L_{ij} does not exceed a specific threshold d_{social} [13]. Figure 5 shows a coherent and non-coherent group structure example.

$$dist(x_{ij}, x_{ij'}) <= d_{social} \tag{8}$$

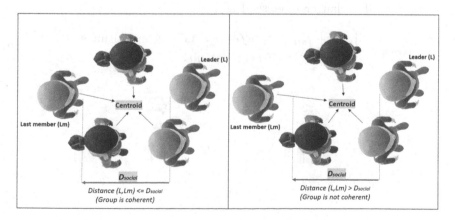

Fig. 5. Coherent group (left) vs. non-coherent group (right)

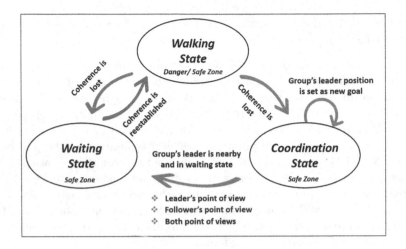

Fig. 6. The movement states of pedestrian groups

Intra-group Interaction: We model the interconnection of group members in three states i.e. walking, waiting and coordination states, illustrated in Fig. 6.

At the beginning of simulation, all group members are in the walking state, in which they walk together using Eq. 6. When the coherence of the respective group G gets lost, group leader L_{ij} switches to the waiting state and waits for the other members, whereas the other members switch to coordination state. In coordination state, the current position of L_{ij} is set as a temporary goal of other members and they stay in this state until they reach to the leader's position. When every member reaches his/her leader, the coherence of G is reestablished and therefore all group members return back to the walking state. However, group members can stay at waiting or coordination states only if they are in safe zone (a place which is safe to wait or coordinate). Pedestrian zones, and mixed zones with no competitive users within any particular group member's field of view can be examples of safe zones. Whereas, overcrowded pedestrian or mixed zones are defined as danger zone, see Fig. 7.

$$f_{group} = \begin{cases} f_{vis} + f_{att}, & \text{if } Safe \text{ zone and not in waiting state} \\ 0, & \text{if } Danger \text{ zone.} \end{cases} \tag{9}$$

Fig. 7. Safe zone SZ (left) vs. danger zone DZ (right)

Pedestrian Group-to-Vehicle Interaction: Interaction between any pedestrian group (G) and vehicle (V) is handled by considering the steps in Fig. 8.

- First V and L decide on their strategy among different alternatives (decelerate, accelerate, deviate) by playing a Stackelberg game, see Subsect. 3.3.
- If G splits, all subgroups of G either follow the strategy chosen by the leader.
- Or all members of the subgroup of G (G_i) which L belongs to, follow the strategy of L and all other subgroups can perform one of the following:
  - all subgroups can follow the strategy of L.
  - or if V and L decide to decelerate and accelerate respectively, then other subgroups can either decelerate or deviate or some subgroups decelerate and others deviate.
  - or if V accelerates and L decelerates, then all other subgroups deviate.

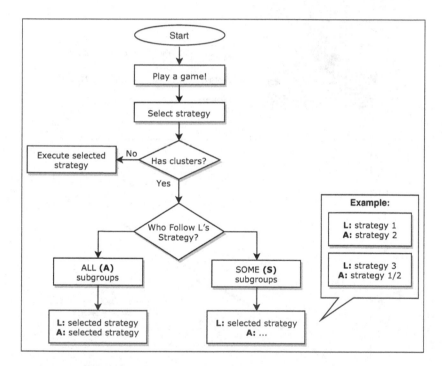

Fig. 8. Execution steps of vehicle-to-pedestrian groups interaction. Here, strategy 1, 2 and 3 represent decelerate, deviate and accelerate respectively.

4   Evaluation

To validate our model, we extract and simulate 20 scenarios from a data set of a shared-space in Hamburg [18], which involve pedestrian groups-to-vehicle conflicts and visualize the difference in real and simulated trajectories and speed of all involved road users. We select one scenario among all those scenarios to illustrate elaborately vehicle-to-group interaction and create two example scenarios, one to capture the interaction between big pedestrian groups and another to show the clustering feature of group. For the first scenario, we compared the real and simulated behaviors and trajectories of the involved road users. However, we could not compare the performance of the last two scenarios as real data is currently missing.

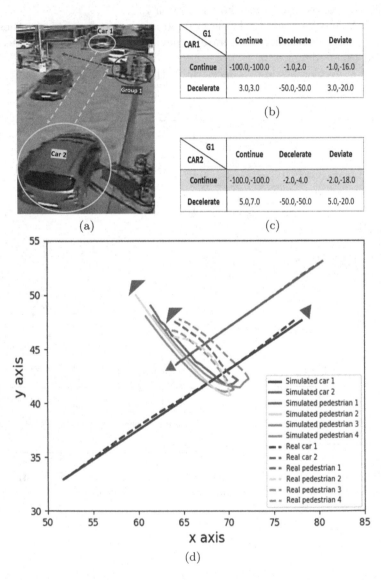

Fig. 9. Scenario 1: A pedestrian group to car interaction. (a) Real scenario snapshot (b) Game matrix: Car1 to Group1, (c) Game matrix: Car2 to Group1 (d) Comparison of the trajectories. The arrows indicate the direction of movement of the road users.

The implementation of our model is performed in Lightjason, a Java-based BDI multi-agent framework [2]. An Intel Core^{TM}i5 processor with 16 GB RAM was used to perform all the simulation runs.

Scenario 1: Here as shown in Fig. 9a an interaction between one pedestrian group (*Group1*) of four members and two cars (*Car1* and *Car2*) is captured. In this scenario, both drivers decelerate to let the group cross the street. In our model, both these car-to-group interactions are handled using game-theoretic model. The game matrices in Fig. 9b and c reflect courtesy of the drivers towards the pedestrian group. In Fig. 9d, the trajectories of both cars are very similar in simulation and real data, but the trajectories of group members in simulation are deviated from the real trajectories, even though the reacted behavior is same.

Scenario 2: This scenario presents the interaction between two pedestrian groups of non-cluster structure. Each group consists of eight members. *Group1* is moving to the southern direction, and *Group2* is moving to the northern direction. As can be seen in Fig. 10, when the groups become very adjacent to each other, they slightly deviate to avoid collision with other. Our model handles this interaction as simple interaction.

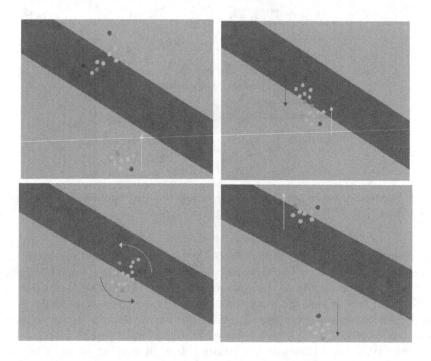

Fig. 10. Scenario 2: Non-clustered group to group interaction

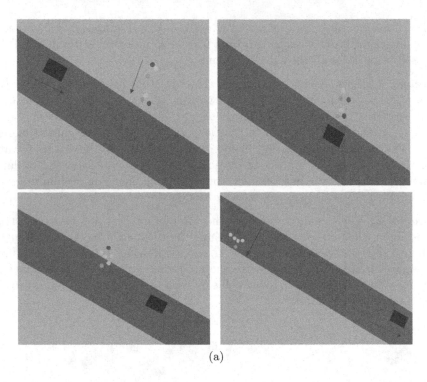

(a)

PED CAR	Continue	Decelerate	Deviate
Continue	-100.0,-100.0	14.0,2.0	14.0,-16.0
Decelerate	3.0,3.0	-50.0,-50.0	3.0,-20.0

(b)

Fig. 11. Scenario 3: Clustered group to car interaction. (a) Simulation snapshots (b) Game matrix

Scenario 3: In Fig. 11, the interaction between a group $Group1$ and a car $Car1$ is represented. In this scenario, $Group1$ consists of six members and two clusters, where each cluster consist of three members. Since the leader L of $Group1$ is in Safe Zone (car is still far away) and lost sight to the last member Lm, L waits for his/her group members to reach L to reestablish the group coherency before crossing the road. Afterwards, when $Car1$ and $Group1$ interact $Car1$ takes priority and continues to go because L is in waiting state. The payoff matrix in Fig. 11b also reflects this factor.

To sum up, in all these scenarios, the pedestrian groups and cars suitably detect and classify the interaction and behave accordingly during simulation.

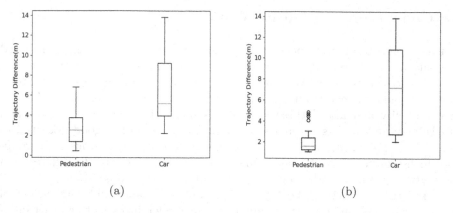

Fig. 12. Trajectory differences of real and simulated road users. (a) Without group model (b) With group model

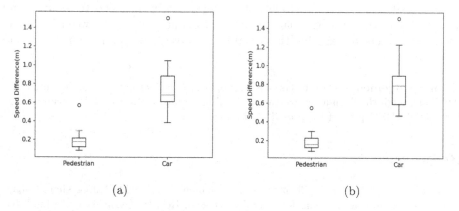

Fig. 13. Speed differences of real and simulated road users. (a) Without group model (b) With group model

Figures 12 and 13 represent the average deviation of trajectories and speed of all road users of our selected 20 scenarios, before and after adding group dynamic to our model. The purpose of this comparison is to see the difference in interaction of car with arbitrary groups (e.g. a group of people waiting to cross a road) vs social groups (e.g. family members or socially related pedestrians walking together). The speed profiles of pedestrians and cars remain almost same after integrating pedestrian group behaviors to our model. Whereas, there is some improvement in terms of pedestrians trajectories (2.89 vs 2.21), but small weakening in cars trajectories (6.58 vs 7.0). Our analysis through simulation modeling says that, social groups take more time compared to arbitrary group to cross in front of the car, hence car has to wait longer causing this fore-mentioned weakening in car trajectory.

5   Conclusion and Future Work

In this paper, we integrate pedestrian group dynamic into our multiagent-based simulation model. Modeling of interactions between social groups and vehicles is our novel contribution. We analyze the interaction among pedestrian group members and also pedestrian group-to-vehicle interaction, classify them in terms of complexity and model these interactions using the social force model and Stackelberg games. First results support that our model has good potential to model both homogeneous (e.g. group-to-group) and heterogeneous (i.e. group-to-vehicle) interactions. We take a first attempt to investigate the difference in interaction of vehicle with arbitrary groups and social groups.

Our future research will be dedicated on calibrating and validating our model parameters, analysing the transferability of our model, investigating other influencing factors such as personal preferences, ages, and gender of road users, which might have impact on their decision making, and integrating explicit interaction among road users into our model. Most importantly, we need to analyze and model larger scenarios with larger numbers of interactions between heterogeneous road users to examine the scalability of proposed interaction types and our simulation model.

Acknowledgements. This research is supported by the German Research Foundation (DFG) through the SocialCars Research Training Group (GRK 1931). We acknowledge the MODIS DFG project for providing datasets.

References

1. Anvari, B., Bell, M.G., Sivakumar, A., Ochieng, W.Y.: Modelling shared space users via rule-based social force model. Transp. Res. Part C Emerg. Technol. **51**, 83–103 (2015)
2. Aschermann, M., Kraus, P., Müller, J.P.: LightJason: a BDI framework inspired by Jason. In: Criado Pacheco, N., Carrascosa, C., Osman, N., Julián Inglada, V. (eds.) EUMAS/AT 2016. LNCS (LNAI), vol. 10207, pp. 58–66. Springer, Cham (2017). https://doi.org/10.1007/978-3-319-59294-7_6. https://lightjason.github.io
3. Bjørnskau, T.: The zebra crossing game-using game theory to explain a discrepancy between road user behaviour and traffic rules. Saf. Sci. **92**, 298–301 (2017)
4. Cheng, H., Sester, M.: Modeling mixed traffic in shared space using LSTM with probability density mapping. In: 2018 21st International Conference on Intelligent Transportation Systems (ITSC), pp. 3898–3904. IEEE (2018)
5. Costa, M.: Interpersonal distances in group walking. J. Nonverbal Behav. **34**(1), 15–26 (2010). https://doi.org/10.1007/s10919-009-0077-y
6. Gettman, D., Head, L.: Surrogate safety measures from traffic simulation models. Transp. Res. Rec. **1840**(1), 104–115 (2003)
7. Helbing, D., Molnar, P.: Social force model for pedestrian dynamics. Phys. Rev. E **51**(5), 4282 (1995)
8. Huang, L., et al.: Social force model-based group behavior simulation in virtual geographic environments. ISPRS Int. J. Geo-Inf. **7**(2), 79 (2018)

9. Johora, F.T., Müller, J.P.: Modeling interactions of multimodal road users in shared spaces. In: 2018 21st International Conference on Intelligent Transportation Systems (ITSC), pp. 3568–3574. IEEE (2018)

10. Kamphuis, A., Overmars, M.H.: Finding paths for coherent groups using clearance. In: Proceedings of the 2004 ACM SIGGRAPH/Eurographics Symposium on Computer Animation, pp. 19–28. Eurographics Association (2004)

11. Kita, H.: A merging-giveway interaction model of cars in a merging section: a game theoretic analysis. Transp. Res. Part A Policy Pract. **33**(3–4), 305–312 (1999)

12. Koefoed-Hansen, A., Brodal, G.S.: Representations for path finding in planar environments. Ph.D. thesis, Citeseer (2012)

13. Kremyzas, A., Jaklin, N., Geraerts, R.: Towards social behavior in virtual-agent navigation. Sci. China Inf. Sci. **59**(11), 1–17 (2016). https://doi.org/10.1007/s11432-016-0074-9

14. Lütteken, N., Zimmermann, M., Bengler, K.J.: Using gamification to motivate human cooperation in a lane-change scenario. In: 2016 IEEE 19th International Conference on Intelligent Transportation Systems (ITSC), pp. 899–906. IEEE (2016)

15. Moussaïd, M., Helbing, D., Theraulaz, G.: How simple rules determine pedestrian behavior and crowd disasters. Proc. Natl. Acad. Sci. **108**(17), 6884–6888 (2011)

16. Moussaïd, M., Perozo, N., Garnier, S., Helbing, D., Theraulaz, G.: The walking behaviour of pedestrian social groups and its impact on crowd dynamics. PLoS ONE **5**(4), e10047 (2010)

17. Nagel, K., Schreckenberg, M.: A cellular automaton model for freeway traffic. J. Phys. I **2**(12), 2221–2229 (1992)

18. Pascucci, F., Rinke, N., Schiermeyer, C., Friedrich, B., Berkhahn, V.: Modeling of shared space with multi-modal traffic using a multi-layer social force approach. Transp. Res. Procedia **10**, 316–326 (2015)

19. Pascucci, F., Rinke, N., Schiermeyer, C., Berkhahn, V., Friedrich, B.: Should I stay or should I go? A discrete choice model for pedestrian-vehicle conflicts in shared space. Technical report (2018)

20. Rinke, N., Schiermeyer, C., Pascucci, F., Berkhahn, V., Friedrich, B.: A multi-layer social force approach to model interactions in shared spaces using collision prediction. Transp. Res. Procedia **25**, 1249–1267 (2017)

21. Schönauer, R.: A microscopic traffic flow model for shared space. Ph.D. thesis, Graz University of Technology (2017)

22. Schönauer, R., Stubenschrott, M., Huang, W., Rudloff, C., Fellendorf, M.: Modeling concepts for mixed traffic: steps toward a microscopic simulation tool for shared space zones. Transp. Res. Rec. **2316**(1), 114–121 (2012)

23. Vizzari, G., Manenti, L., Crociani, L.: Adaptive pedestrian behaviour for the preservation of group cohesion. Complex Adapt. Syst. Model. **1**(1), 7 (2013). https://doi.org/10.1186/2194-3206-1-7

ANNO: A Time Series Annotation Tool to Evaluate Event Detection Algorithms

Jana Huchtkoetter[(✉)], Andreas Reinhardt, and Sakif Hossain

Department of Informatics, Technische Universität Clausthal,
Julius-Albert-Str. 4, 38768 Clausthal-Zellerfeld, Germany
{jana.huchtkoetter,sakif.hossain}@tu-clausthal.de, reinhardt@ieee.org

Abstract. The research field of energy analytics is concerned with the collection and processing of data related to electrical power generation and consumption. Electricity consumption data can reveal information pertaining to the nature of underlying appliances, their mode of operation, and many other aspects. Sudden load changes, so-called *events*, constitute the principal source of information in such time series data, thus their reliable detection and interpretation is a prerequisite for accurate energy analytics. The development of event detection algorithms is, however, hampered due to the unavailability of comprehensive data sets that feature energy consumption time series with corresponding event annotations. We hence present ANNO, a tool to provide annotations to time series consumption data in a supervised fashion and use them for the development of energy analytics algorithms, in this work.

Keywords: Load signature analysis · Supervised data set annotation

1 Introduction and Motivation

Electrical load signatures are the principal sources of information in energy analytics. Their analysis has sparked a strong worldwide research interest [25]. Practical use cases based on load signature analysis include the identification of operating appliances [24], their predictive maintenance [6], and the realization of smart homes [17]. The development and improvement of algorithms for load signature analysis is, however, strongly reliant on the availability of data collected from actual real-world dwellings. To this end, a number of energy data sets have been published (e.g., REDD [15], Tracebase [21], and BLOND [16]). While these data sets have been used in numerous research works on energy analytics, a commonality that hampers their widespread use is the absence of annotations. This makes the validation of newly developed algorithms difficult. Another option to test and evaluate algorithms would be the use of artificial data, created by simulating the usage of devices in a dwelling [7]. While such artificial data holds many possibilities to make evaluation and testing of algorithms easier [8], its creation requires annotated device models and its validation against real world data is only possible with ground truth data which provides necessary labels. Data sets

© Springer Nature Switzerland AG 2020
N. Gunkelmann and M. Baum (Eds.): SimScience 2019, CCIS 1199, pp. 70–87, 2020.
https://doi.org/10.1007/978-3-030-45718-1_5

with annotations (e.g., BLUED [1]) partially overcome these limitations. There are still some disadvantages with these data sets, however. For example, the number of contained traces is often insufficient to generalize the results obtained through their analysis. Furthermore only very few data sets offer complete annotations of all contained events. This hampers the investigation of new aspects, e.g., the impact of differing temporal resolutions or geographical locations on the performance of energy analytics algorithms. Additionally, simulation has lately emerged as a way to create sufficient amounts of realistic data for algorithm development. While the existing data provides possibilities to generate artificial traces, a greater number of annotated traces from different locations enables the possibility to validate this artificially created data. Another advantage of fully annotated data is the easier and more realistic creation of simulation models.

We follow the concept of making comprehensive data sets usable within the scope of energy analytics research by adding annotations to them. To this end, we present our annotation tool called ANNO, which allows for the visual review of time series data, and features an intuitive and rapid work flow to mark events and enter annotations regarding their nature. Aside from the ability to annotate data, the tool will also offer the possibility to review already made annotations, which can be used to expand on existing annotation or evaluate the results of e.g. event detection algorithms.

The paper is structured as follows. We begin by exploring related work in Sect. 2, with an emphasis on existing data sets, energy analytics algorithms, and applications. In Sect. 3 we illustrate the design decisions and requirements behind ANNO. In Sect. 4, we explain its mode of operation as well as its output format, as well as the key challenges that had to be tackled during its implementation. A proof of concept and small usability test are presented in Sect. 5, before we finish with conclusions and an outlook in Sect. 6.

2   Related Work

The field of energy data analysis has seen many advances in the past decade. This is also well-aligned with the advances and regulations to protect the environment and avert climate change, as load monitoring, and the constant feedback on energy usage have been proven to enable higher energy savings than indirect feedback through billing [2,9].

2.1   Energy Data Collection

To provide consumers with this information there are two main approaches: Intrusive load monitoring (sometimes also called distributed sensing), and non-intrusive load monitoring (NILM, also known as single-point sensing) [25]. Intrusive load monitoring aims at gathering information about the running appliances by outfitting all or at least multiple power outlets with measuring devices. The measurements of these sensors are then centrally collected and processed. This process bears the disadvantage of being difficult and time-consuming to install, as well as costly due to the great number of sensing devices needed. Non-intrusive load monitoring, tries to overcome this need by just requiring a single point of sensing, usually at the electricity meter [11]. This way, electrical consumption data is collected for the whole house, and the task of NILM algorithms is to correctly disaggregate this data into the individual devices that caused the energy consumption. Because of the economical advantages in comparison to intrusive load monitoring, NILM promises a wide applicability, and has in consequence seen extensive research and innovation in the past three decades [11].

NILM is based on three steps: During the *data acquisition* the electrical consumption data is collected in order to be provided to the following steps. The sampling frequency ranges from a few Hz or lower to frequencies on the order of several kHz. The collected data is provided to the *feature extraction* step. Computed features widely vary, depending on the NILM algorithm, and can be divided into steady-state and transitional features [22]. Steady-state features are collected during periods in which the electrical load is constant. This is usually the case when operated appliances do not experience any state changes. Transitional features describe the behaviour of electrical measurements during the transition between two states. The detection of sudden changes (so called events) is also part of the feature extraction step. Third and lastly, the *load identification* step targets to identify appliances from the features computed in the previous steps. In case an event detection was performed, the temporal offsets of detected events are usually taken into consideration.

2.2   Data Sets for Energy Analytics

A lot of research has been centred around the development and evaluation of efficient event detection and load identification algorithms. To test such algorithms a set of ground truth data is required. Such data is compromised of the consumption data, on which the algorithm runs, and additional information which makes the measurement of the algorithm's accuracy possible. In the case of electric consumption data this usually means that the data is annotated with information when which device was turned on or off. Depending on the algorithm and the research goal, a variety of additional information can be included. The acquisition of such data can be a time consuming and cost intensive process. Ground truth data needs to be collected in parallel to the consumption data and reviewed afterwards. Consumption data should, to capture behavioural patterns and changes, be collected over the span of several months. Furthermore, the collection of data from more than one dwelling adds variety, but also complexity [14].

These considerations in combination with the need for publicly available data on which algorithms can be compared have led different researchers to the acquisition, review, and publication of data sets. The different data sets provide data with different sampling frequencies, length, and scope. The comparison alone already poses some difficulties, as the reason for data collection and the conditions under which the data sets were created and published are differing [14].

The easiest possibility to divide data sets is the frequency at which measurements were taken, ranging from several kHz down to fractions of 1 Hz. In this paper we illustrate some of the data sets which contain data sampled at 1 Hz or more often. In order to give the reader a feeling for the heterogeneity of data sets a table was created, containing selected features (see Table 1). Data set properties showcasing the highly heterogeneous nature are listed, namely the existence of event annotation, sampling levels (whole-house, circuit, or plug level), number of dwellings, number of applications, application classes in the data set, and additional information included. It needs to be noted, that albeit a data set which contains plug level data annotated with the device connected to this plug may be seen as annotated, this table will mark those as *partially annotated*. This is due to the fact that training an event detection algorithm on the associated whole-house or circuit data would still require some work in inspecting and noting down event times. Unless otherwise noted, the data sets were collected from residential dwellings.

As already stated and visible from Table 1, only few data sets offering annotations exist. A short look into BLUED's [1] information about the annotation reveals a reason: All activities needed to be documented while the collection was going on. After the collection further visual inspection of the consumption data was conducted to confirm event timings before compiling the list of events. This excessive amount of work justifies the missing annotations in other data sets.

2.3 Simulation of Electrical Consumption Data

The differences and variations in data sets in combination with the requirement for well-known and annotated data for algorithm development has led to the creation of simulation tools for electrical consumption. Based on the same model for device model generation [3], two different simulation frameworks have been developed [7,8]. While they differ in their possibilities, both provide the means to create high amounts of realistic artificial data.

But even these approaches struggle with the lack of well annotated real world data. On the one side device model generation requires annotated data. On the other side, when validating simulation data, only fully annotated data sets can be used. Therefore, simulation data can only show its validity concerning the few annotated data sets and would therefore highly benefit from the possibility to annotate as of yet unannotated data sets.

2.4 Related Approaches

The ANNO tool is not the first tool created to annotate existing data sets and therefore overcome some of the mentioned challenges. Two publications from the research field of energy informatics handle the same idea. In [23] the authors aimed at annotating the Pecan Street data set through an expert crowd sourcing approach for the annotation. To achieve this goal a collaborative online framework has been created. It chooses data snippets from the data set to be annotated and features a leader board to increase the user motivation and engagement.

Table 1. Comparative table of data sets

Name	Sampling rate	Places	Appliances	Duration	File format
BLOND-250 [16]	250 kHz	1	-	50 days	hdf5
BLUED [1]	12 kHz	1	ca. 50	7 days	csv
COOLL [20]	100 kHz	1	42	-	flac
GREEND [18]	1 Hz	9	48	ca. 365 days	csv
iAWE [5]	1 Hz, 20 Hz	1	> 11	ca 120 days	csv/hdf5
PLAID [10]	30 kHz	1	> 1800	-	csv
REDD [15]	15 kHz, 1 Hz	5	92	> 2 weeks	text file
Smart* [4]	1 Hz or less	3	-	6 months	wind energy
Tracebase [21]	1Hz	ca. 20	122	-	csv
UK-DALE [13]	16 kHz, 1/6 Hz	3	> 75	4.3 years	csv/hdf5

Name	Sampling levels	Annotated	Notes and further information
BLOND-250 [16]	Whole-house, appliance	Partially	Office building
BLUED [1]	Whole-house	Yes	Often used for benchmarking
COOLL [20]	Appliance	Partially	Turn-on inrush measurements; measured in laboratory
GREEND [18]	Appliance	No	In Italy and Austria
iAWE [5]	Whole-house, circuit, appliance	No	Water and ambient information; measured in India regards specialities (e.g., outages)
PLAID [10]	Appliance	Yes	1 s start-up and steady-state; laboratory sampled
REDD [15]	Whole-house, appliance	No	Additional geographical features
Smart* [4]	Whole-house, circuit,appliance	csv	Environmental and occupational data, created solar & wind energy
Tracebase [21]	Appliance	Partially	Granularity of individual devices; offers device information for disaggregation training
UK-DALE [13]	Whole-house, appliance	No	Additional lower sampled set available

Unfortunately the website containing this tool is not available any more and therefore of no use for the current task. It is furthermore designed to work best with a large crowd of contributors, whereas ANNO is meant to enable single persons to annotate data. Pereira et al. [19] argue that the lack of labelled data sets is caused by the high amount of work required in inspecting measured data for labelling The created semi-automatic labelling tool provides the user with an overview of the data and implements a detection algorithm which then only needs to be supervised. It furthermore reduces the amount of data by removing data with similar values through steady-state detection. The created labels are stored in a database, as well as the data to be annotated. Unfortunately, the tool only offers the possibility to label an event as an event, without providing the possibility to introduce further meta data.

As anomalies are not only of relevance in energy informatics, but also in other fields, a few other tools exist to label time sequence data. For example [12] provides a short list, naming Curve[1], TagAnomaly[2], time-series-annotator[3] and WDK[4]. While all of these tool offer slightly different possibilities, none of them are able to annotate additional meta data as provided by BLUED and required for NILM research.

We have therefore decided to create a tool aimed at labelling already existing data sets: ANNO. The tool is better suited for this approach as it utilizes the already existing data files and aims at making the inclusion of data sets as easy as possible, while providing interfaces to add meta data aside from the event location.

3   Design Decisions

3.1   Output File Format

One key design decision to make is the format in which event annotations are stored. BLUED [1] is the data set with the most complete annotations and has furthermore been used frequently as a de-facto standard for algorithm benchmarking. It was therefore decided that ANNO's output format needs to be compatible with BLUED's event annotation format. BLUED lists a timestamp, a label, and the mains phase on which each event occured. The labels are device IDs indicating which (type of) device caused the event. The file format is commaseparated and, as such, can be used on a wide range of systems. Although the information collected in BLUED already enables the training and evaluation of many energy analytics algorithms, there are further relevant features that a user might want to choose, such as a mark for the event type (turning on, turning off, or unknown), the start of the corresponding transition, the length of the transition and a free comment field. This field can be used to note further information about the event, enabling users to augment data sets exactly to their needs. A header and example data can be found in Table 2.

3.2   Trace Review Interface

The inspection of time series data of electrical loads bears some challenges. This is due to the fact that throughout an average day several intervals with very low or no activity are present, some of which can span over multiple hours. Depending on the type of building this might either be during the day, when most tenants are at work or in school, or during the early morning and evening for business buildings, as well as during the night. During these periods inspection can ideally occur over a wide time window, whereas during higher activity periods more

---

[1] https://github.com/baidu/Curve.
[2] https://github.com/Microsoft/TagAnomaly.
[3] https://github.com/CrowdCurio/time-series-annotator.
[4] https://github.com/avenix/WDK.

Table 2. Excerpt from ANNO's annotation output

Time	Label	Phase	Power	Marker	Trans. start	Trans. end	Remarks
10/20/2011 17:27:04.430916	spike	B	356.3	NA	NA	NA	spike

events take place and a smaller temporal granularity may be advantageous for inspection. This need is strengthened when taking the possibility of overlapping events into account which demands for even higher granularity. For ANNO we decided to approach this issue from two directions. First, an overview over the whole day is available to ease the visual inspection and identify high activity areas. To this end, ANNO embeds two plots into the user interface as can be seen in Fig. 1. The lower, smaller plot is an overview of the whole data file which is going to be inspected. It serves as an orientation and will further be called *navigation plot*. A user can easily identify areas in the graph which include higher activity or may already be able to make out coarse events. They can then use this graph to zoom into the area of interest. The chosen area is displayed at a finer time granularity in the upper graph. This graph is displayed larger to enable exact visual inspection of areas of interest and is therefore called *inspection plot*.

To further be able to adapt to the different length of transitions, frequency of events, and user preferences, the temporal resolution of the upper graph can be changed. To do this, the user can enter the time length that should be displayed in the window. The main work of annotating found events is done in the inspection plot. The features mentioned concerning the output format in Subsect. 3.2 need to be annotated to the plot during the inspection. This annotation is done mostly through key strokes, such that a rapid work flow is possible. This work flow starts by annotating an event. Three different key bindings are defined for the user to directly mark a change as either an arbitrary, an *on*, or an *off* event. The event types can be differentiated through their colours. As soon as an event marker is set through the user pressing a key, the corresponding event goes into edit mode. During the editing process, it is furthermore possible to change the other event properties. To this end, a transition start and a transition end time can be added. If the user does not add them, the transition start and end time will be noted as unknown. The features that are not directly connected to a time and amplitude on the graph (and therefore cannot be extracted automatically from the marked events or be marked themselves) are noted down in a separate, adjoining area. This area holds input fields for the label, remark, and phase. When the user is satisfied with the created event he can finish the editing, after which the process can start again with the next event. Alternatively marking the next event also saves the previous one and starts a new editing process.

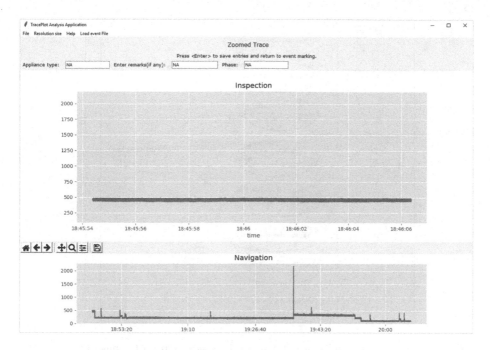

Fig. 1. User interface of ANNO with loaded consumption data.

3.3   Importing Data Sets

Regardless of the usefulness or importance of editing possibilities, they can only be meaningful if they can be applied to useful data sets. Therefore the possibility to integrate arbitrary data sets into the tool was deemed highly important. Albeit there are many different collections of load data, there is no universally agreed upon format they are presented in. Some data sets, like BLUED [1] or GREEND [18], apply the commonly used .csv format. This format can be read easily by many programming languages, but stores the data uncompressed. As time series data can require high amounts of storage memory, especially when recorded using high sampling rates, this can be disadvantageous. To counter this, other data sets have adopted compressed file formats, like the .hdf5 format, or formats designed for sequential (albeit not electrical) data, like the .flac format [20].

Besides the file format differences, there are also differences in the data collected and its organization. Every data set contains a multitude of measurements, differing in many ways, from the aspects measured (i.e., current, voltage, power, etc.) to the sampling rate applied. This was already demonstrated in Table 1. Because of this diversity, the tool must offer a modular way to load data sets. To ensure the easy inclusion of new data sets, ANNO was designed to handle the data set loading and graphical user interface (GUI) integration with as few interfaces as possible. All necessary information was supposed to be given at one location in the code, which just needs to include a selection possibility for the new data set and should return the values to be displayed. The further GUI

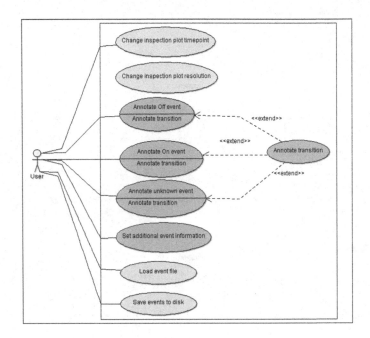

Fig. 2. Use case diagram for ANNO's main interface, after a file has been loaded.

integration are handled by the tool as autonomously as possible. This guarantees that a developer only knowing how to load a new data set can integrate it, without having to learn about the tool's further internal workings.

4   Operation of ANNO

In the following section we will present ANNO as a tool and give information about its usage. To better understand how ANNO can be used, the use case diagram Fig. 2 was created.

ANNO's usage always starts with loading a datafile and choosing which data type it has (i.e., which data set it belongs to) and which aspects should be used for x- and y-axis values. After this choice the file will be loaded and displayed in the main GUI (as seen in Fig. 1). From this point three kinds of operations are available to the user. They can either adapt the data representation, by changing the time resolution of the *inspection plot* or selecting a new time point to be displayed in the *inspection plot*, load or save event files or annotate the data. The annotation process comprises multiple possible use cases, namely the annotation of an *on, off,* or *unknown* event which than enables the annotation of the corresponding transition and the possibility to set the additional event information concerning the phase, remark, etc. An example of a freshly loaded file and the ANNO user interface is given in Fig. 1. Furthermore Table 3 contains all implemented key bindings and a short explanation of their purpose. A plot with marked events can be found in Fig. 3. The additional meta data that cannot be marked on the graph directly is provided in separate text entries which are visible above the *inspection plot*.

To fulfil these requirements ANNO was created in the Python programming language. Python is often used in conjunction with big data analysis and has some very useful libraries for handling it. The tool uses three of them. *NumPy* is regularly used in scientific computing and offers data structures besides fast mathematical and numerical functions[5]. For ANNO it is mostly used to run calculations on the bigger data structures. Another library for data analysis is the Python Data Analysis Library (pandas)[6], which offers data structures and analysis tools. ANNO uses it to efficiently handle time series data, as explained in Subsect. 4.2. To display the data to be annotated the pyplot module was used, which provides a Matplotlib[7] interface.

4.1 Loading and Saving

When using ANNO, loading and saving files is an important part of the work. While the saving has a defined output format, the differing possible input formats make loading the part of the tool most likely requiring adaptations. Therefore, it is handled in a source code file dedicated to this use only. The file holds all the information which needs to be extended when a new data set is to be included, such as the interfaces for x- and y-value data. This barres the including user from the need to inspect or adapt code that is not directly related to the data set they want to include.

A feature for the better usability of ANNO is the possibility to save and load event files. The load dialogue can be accessed from the menu bar. The most straightforward option in this context is the loading of event files annotated through ANNO. After choosing a file it will be loaded, displaying the previously entered annotations as markers. Being able to reload already made annotations facilitates the annotation of long data files. However, the possibility to display previously entered event annotations holds even more potential. Therefore ANNO is not only able to load event files in its output format, but also in the BLUED format, as well as simple timestamps. This enabled the loading of ground truth files from BLUED, as well as events found by an algorithm. Thus ANNO features an easy possibility to visually check event detection results.

4.2 Implementation Challenges

The creation of ANNO was faced with a set of challenges. Many of those arose from the specific requirements of electric load data and the different types of data sets the tool must be able to support.

Table 3. Key controls for the ANNO tool.

Control	Plot	Function
Mouse click	Navigation	Displays the clicked area in the inspection plot
Menu point *Resolution size*	-	Change inspection plot resolution to the entered time length (in seconds)
Menu point *Load event file*	-	Opens a file selection dialogue and loads the events from the selected file
W	Inspection	Sets a mark for an *on* event on the graph, taking the x-value from the current mouse pointer location
A	Inspection	Sets a mark for an *off* event on the graph, taking the x-value from the current mouse pointer location
D	Inspection	Sets a mark for an *unknown* event on the graph, taking the x-value from the current mouse pointer location
Q	Inspection	Marks the start of the transition belonging to the currently edited event
E	Inspection	Marks the start of the transition belonging to the currently edited event
Z	Inspection	Deletes the event and corresponding transitions nearest to the mouse pointer
S	Inspection	Saves the currently annotated events to the disk

The data loading posed a steep challenge. Many data sets apply high sampling rates. Those sampling rates are needed to fulfil the requirements for algorithm development (see also Sect. 2). Depending on the chosen sampling rate, the data size in a window of just two minutes can range up to 200 MB, e.g, for the BLOND data set [16]. On top of that, to enable a meaningful visual inspection, longer time frames are required to provide an overview. The loading of meaningful time frames, containing hours or even a whole day of data, was taking exceeding amounts of time.

In the end a property of the data was exploited to enable faster loading. The measurements in question for visual inspection, may it be voltage, current, or some other feature, are always correlated with either a time or a comparable index. This enabled the use of *Series* from pandas. *pandas.Series* is specifically designed to save time series data by offering an one-dimensional array with axis labels[8]. Using standard methods would lead to time and annotation data only

---

[8] For further information consult https://pandas.pydata.org/pandas-docs/stable/reference/api/pandas.Series.html.

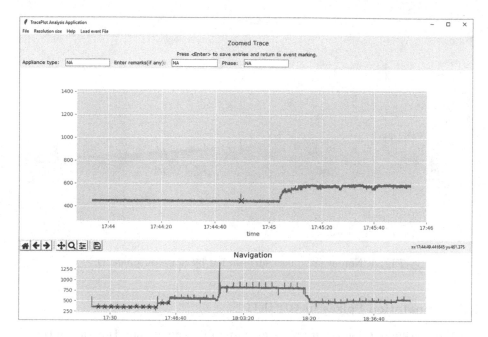

Fig. 3. ANNO's user interface during the proof of concept, marking of non-annotated spikes in BLUED.

being loosely correlated. *Series* objects overcome these problems by offering a data structure that can be accessed either through the data's position or through label-based indices, e.g., time data. In the case of ANNO, *Series* objects not only helped cut loading times, but furthermore made accessing the data easier. It needs to be mentioned that albeit loading times could be substantially decreased, a longer starting time can still occur when dealing with high volumes of data. The long starting time was deemed preferable compared to loading times during the tools usage, as those would be potentially slowing down users and thus hindering work to a greater degree.

5 Proof of Concept

To prove the efficiency of our tool, we provide three different proofs of concept. First the annotation of irregular spikes in BLUED was performed to be able to better evaluate energy analytics algorithm regarding their detections. Second the possibility to annotate so far unannotated data sets were tested. This was done in comparison to one tool used for anomaly annotation to give a feeling how well ANNO fits its main purpose. Lastly a usability test with one user was contracted to see if the general flow of the tool is working and identify points which need further development.

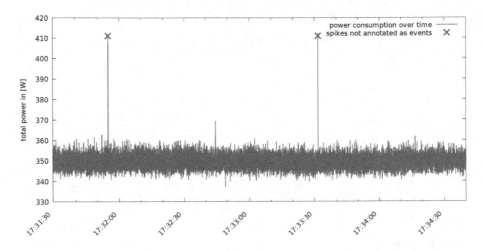

Fig. 4. Non-event spikes surrounded by noise on phase B.

5.1 Annotation Spikes in BLUED

For the marking of non-event spikes a complete manual inspection of the traces needed to be performed. The spikes can easily be identified in the data, as they occur at very uniform distances. They are furthermore very short but of larger amplitude than the surrounding noise, as can be seen in Fig. 4. The effect of the annotation underlines its necessity. Even in a short time many spikes occur, for example an 80 min stretch of data already containing 22 events was enhanced to contain the original 22 events and an additional 47 events which were marked as spikes. An excerpt of the work can be seen in Fig. 4. During the manual trace inspection, it was furthermore noted that, albeit BLUED does define events as having at least an amplitude of 30 W and a length of 5 s, not all points annotated as events follow this rule. This is especially remarkable with the signature of the printer included in the data set. Its signature is displayed in Fig. 5. 46 events had been registered in this time slot, many of which are less than 5 s apart.

Although the tool could have corrected this deviation from the definition, the annotations were left as they were, such that an algorithm finding these events could still be validated.

5.2 Comparison with Another Tool

To enable a fair comparison, it was decided to conduct a proof of concept on completely new and unannotated data. For this purpose a short set of measurements of 1.5 h was taken at TU Clausthal. The choice of tool to which ANNO should be compared was based on the tools mentioned in Sect. 2. It was decided to use *TagAnomaly*[9], a tool developed by Microsoft for anomaly detection and

---

[9] https://github.com/Microsoft/TagAnomaly.

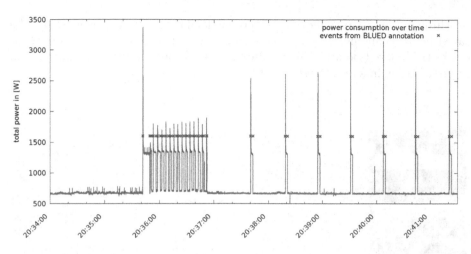

Fig. 5. Printer signature and events annotated by BLUED. Some events between 20:36:00 and 20:37:00 do not adhere to the 5 s length stated in the BLUED paper [1].

labelling. The tool is designed to be able to handle multiple categories, so it could be used to concurrently inspect and annotate voltage, current and power.

For the comparison of the tools, only power data was analysed and labelled. Initially the whole 1.5 h were supposed to be analysed. Unfortunately *TagAnomaly* was unable to handle a file of this size. Therefore, only the first 30 min were labelled with both tools.

The process of preparing and loading a file took a similar amount of time for *TagAnomaly* and *ANNO*. As *TagAnomaly* requires a special file format a short script was written to transform from the measured file format to the new one, while *ANNO* needed to include the data set format as a new format. The file loading in both tools was also comparable. As the user conducting the test was already familiar with *ANNO*, adaptation time to the new user interface was not taken into account.

The first differences became apparent during the annotation. While *TagAnomaly* is intended to label time series data, it is not supposed to label single points. Rather all points belonging to one anomaly are marked as abnormal. This meant that, for setting one point in time as an event, a later inspection of the selected points would be necessary. It furthermore indicated that ideally as few points as possible should be selected with the tool. To ensure this, the view needed to be adapted. *TagAnomaly* offers the possibility to use a slider to restrict the graph to a section of the data. The selection is done through the sliders position on a time line of the data. The section in between the sliders is displayed. Therefore the data can be limited by moving the sliders on the time line. Unfortunately, every correction of a slider triggered a new loading process, which took as long as the initial loading. This made an exact selection of a section of interest very time consuming and tedious. As soon as a selection was set through the sliders, annotations were made. These annotations needed to be saved before changing the selection again, as any change in the selection would delete the marked points. This resulted in a total of 60 min for the annotation of just three points of interest (Fig. 6).

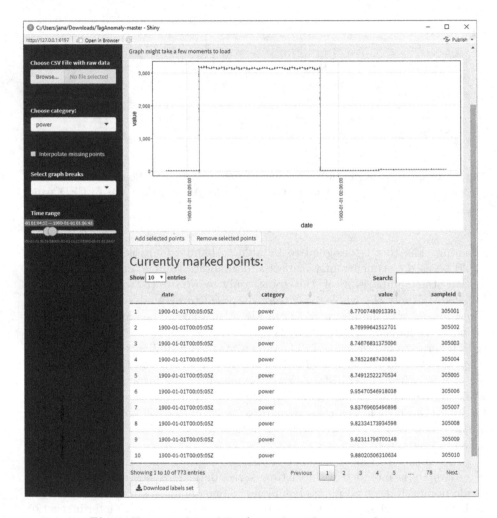

Fig. 6. User interface of TagAnomaly with annotated events

The usage of *ANNO* was a lot faster. This is mostly due to the fact, that *ANNO* is designed to handle a lot of data and not induce long loading times after initial loading. Furthermore *ANNO* differentiates from *TagAnomaly* and other found tools by offering the possibility to enhance the simple annotation through labelling the kind of event and external circumstances, like causing devices. The comparison has proven the utility and advantages of ANNO over other, already existing tools for comparable applications.

5.3   User Test

To check if the tool meets its requirements to enable a fast work flow and good usability, a user test was conducted. The user was tasked to load a file, given the necessary specifications regarding the file type and which properties to plot and was asked to annotate events. The user was already proficient in the domain of NILM data.

Initially the tool was remarked to miss some guidance. The initial loading time before the annotation window is loaded was marked as confusing, as well as the missing guidance which plot is supposed to be used in which way. This was worsened as the tool's usage could only be understood after finding and reading the help. It was furthermore remarked that mouse operations would have been a more intuitive choice for event marking.

Apart from the guidance and as soon as the help was read and understood the tool was easy to use. While getting to know the tool and the way to input data took some time, the idea of making annotations through different key presses showed itself to be a good way to enable fast annotations. Especially the choice of interesting areas from the *navigation plot* to then annotate them in the *inspection plot*, was used intuitively and enabled precise navigation.

Overall the test proved that the tool is going into the right direction to enable a fast and easy way to annotate so far unannotated data. The missing guidance found during the user test will be worked on, such that the time to familiarize with the tool can be reduced.

6   Conclusions and Outlook

The developed tool, ANNO, eases annotations of electrical consumption data. Through the usage of libraries specializing in the analysis and storage of time series data, it was possible to enable fast loading times. This ensures the user can be working with a plot of the whole data for navigation and a smaller plot for inspection of possible events and their marking. In combination with a defined integration point for new data sets, ANNO offers the possibility to annotate any kind of data set. This enables users to not only annotate newly collected data, but also to augment existing public data sets. ANNO's functionality was shown in a set of proof of concepts, proving that ANNO can be used on existing as well as on newly measured data sets. It was furthermore demonstrated that ANNO not only offers more complex and complete annotations than already existing tools, but is also better suited for its purpose by taking the high amounts of data and corresponding loading times into account. The user test showed the usability of the tool for unfamiliar users, albeit some short comings regarding guidance in the first steps were found. These will be tackled in further development.

In the future further improvements to the ANNO tool are possible. A good first step would be the better support during the annotation of events. One of the most intuitive steps towards better annotations is the possibility to find local minima and maxima by the tool. Even with high display resolution the human accuracy might not be sufficient to correctly point out change points in data sampled at rates of multiple kHz. An automatic detection of extremal points would ensure that the chosen events correctly outline the minimal or maximal point of change.

References

1. Anderson, K., Filip, A., Benítez, D., Carlson, D., Rowe, A., Bergés, M.: BLUED: a fully labeled public dataset for event-based nonintrusive load monitoring research. In: Proceedings of the 2nd Workshop on Data Mining Applications in Sustainability (SustKDD) (2011)
2. Armel, K.C., Gupta, A., Shrimali, G., Albert, A.: Is disaggregation the holy grail of energy efficiency? The case of electricity. Energy Policy **52**(1), 213–234 (2013)
3. Barker, S., Kalra, S., Irwin, D., Shenoy, P.: Empirical characterization and modeling of electrical loads in smart homes, pp. 1–10 (2013)
4. Barker, S., Mishra, A., Irwin, D., Cecchet, E., Shenoy, P., Albrecht, J.: Smart*: an open data set and tools for enabling research in sustainable homes. In: Proceedings of the Workshop on Data Mining Applications in Sustainability (SustKDD) (2012)
5. Batra, N., Gulati, M., Singh, A., Srivastava, M.B.: It's different: insights into home energy consumption in India. In: Proceedings of the 5th ACM Workshop on Embedded Systems For Energy-Efficient Buildings (BuildSys) (2013)
6. Bonaldi, E., de Lacerda de Oliveira, L., Borges da Silva, J., Lambert-Torres, G., Borges da Silva, L.: Predictive maintenance by electrical signature analysis to induction motors. In: Esteves Araújo, R. (ed.) Induction Motors - Modelling and Control. IntechOpen (2012)
7. Buneeva, N., Reinhardt, A.: AMBAL: realistic load signature generation for load disaggregation performance evaluation. In: 2017 IEEE International Conference on Smart Grid Communications (SmartGridComm), pp. 443–448 (2017)
8. Chen, D., Irwin, D.E., Shenoy, P.J.: SmartSim: a device-accurate smart home simulator for energy analytics. In: 2016 IEEE International Conference on Smart Grid Communications (SmartGridComm), pp. 686–692 (2016)
9. Ehrhardt-Martinez, K., Donnelly, K., Laitner, J.: Advanced Metering Initiatives and Residential Feedback Programs: a Meta-Review for Household Electricity-Saving Opportunities. American Council for an Energy-Efficient Economy (2010)
10. Gao, J., Giri, S., Kara, E.C., Bergés, M.: PLAID: a public dataset of high-resolution electrical appliance measurements for load identification research: demo abstract. In: Proceedings of the 1st ACM Conference on Embedded Systems for Energy-Efficient Buildings (BuildSys) (2014)
11. Hart, G.W.: Nonintrusive appliance load monitoring. Proc. IEEE **80**(12), 1870–1891 (1992)
12. Heartex Inc.: A Curated List of Awesome Data Labeling Tools. https://github.com/heartexlabs/awesome-data-labeling
13. Kelly, J., Knottenbelt, W.: The UK-DALE dataset: domestic appliance-level electricity demand and whole-house demand from five UK homes. Sci. Data **2**(150007) (2015). http://jack-kelly.com/data/
14. Klemenjak, C., Reinhardt, A., Pereira, L., Berges, M., Makonin, S., Elmenreich, W.: Electricity consumption data sets: pitfalls and opportunities. In: Proceedings of the 6th ACM International Conference on Systems for Energy-Efficient Buildings, Cities, and Transportation (BuildSys), pp. 159–162 (2019)

15. Kolter, J.Z., Johnson, M.J.: REDD: a public data set for energy disaggregation research. In: Proceedings of the Workshop on Data Mining Applications in Sustainability (SustKDD) (2011)

16. Kriechbaumer, T., Jacobsen, H.A.: BLOND, a building-level office environment dataset of typical electrical appliances. Sci. Data **5**, 180048 (2018)

17. Masoodian, M., André, E., Kugler, M., Reinhart, F., Rogers, B., Schlieper, K.: USEM: a ubiquitous smart energy management system for residential homes. Int. J. Adv. Intell. Syst. **7**(3&4), 519–532 (2014)

18. Monacchi, A., Egarter, D., Elmenreich, W., D'Alessandro, S., Tonello, A.M.: GREEND: an energy consumption dataset of households in Italy and Austria. In: Proceedings of the 5th IEEE International Conference on Smart Grid Communications (SmartGridComm) (2014)

19. Pereira, L., Ribeiro, M., Nunes, N.: Engineering and deploying a hardware and software platform to collect and label non-intrusive load monitoring datasets. In: Proceedings of the 5th IFIP Conference on Sustainable Internet and ICT for Sustainability (SustainIT), pp. 1–9 (2017)

20. Picon, T., Nait Meziane, M., Ravier, P., Lamarque, G., Novello, C., Le Bunetel, J.C., Raingeaud, Y.: COOLL: Controlled On/Off Loads Library, a Public Dataset of High-Sampled Electrical Signals for Appliance Identification. arXiv preprint arXiv:1611.05803 [cs.OH] (2016)

21. Reinhardt, A., et al.: On the accuracy of appliance identification based on distributed load metering data. In: Proceedings of the 2nd IFIP Conference on Sustainable Internet and ICT for Sustainability (SustainIT), pp. 1–9 (2012)

22. Sadeghianpourhamami, N., Ruyssinck, J., Deschrijver, D., Dhaene, T., Develder, C.: Comprehensive feature selection for appliance classification in NILM. Energy Build. **151**, 98–106 (2017)

23. Sandlin, H.A., Kurniawan Wijaya, T., Aberer, K., Nunes, N.: A collaborative framework for annotating energy datasets. In: Proceedings of the 2015 Workshop for Sustainable Development at the 2015 IEEE International Conference on Big Data (BigData) (2015)

24. Weiss, M., Helfenstein, A., Mattern, F., Staake, T.: Leveraging smart meter data to recognize home appliances. In: Proceedings of the IEEE International Conference on Pervasive Computing and Communications (PerCom) (2012)

25. Zoha, A., Gluhak, A., Imran, M.A., Rajasegarar, S.: Non-intrusive load monitoring approaches for disaggregated energy sensing: a survey. MDPI Sens. **12**, 16838–16866 (2012)

Simulation of Materials: Self-Organized and Porous Structures

Vibration Frequency Spectrum
of Water-Filled Porous Silica Investigated
by Molecular Dynamics Simulation

Yudi Rosandi$^{(\boxtimes)}$ (iD) and Gheo R. Fauzi

Department of Geophysics, Universitas Padjadjaran, Sumedang 45363, Indonesia
rosandi@geophys.unpad.ac.id

Abstract. Using Molecular Dynamics (MD) method we evaluated the
vibration characteristics of nanoporous silica material. The purpose of
this work is to observe the effect of water filling inside the nanopores
to the vibration property of silica-water compound system. The vibration frequency spectrum was obtained through the calculation of the
velocity auto-correlation function during MD simulation execution. The
interaction of atoms in the target was modeled using Reax Force Field.
We varied the water density in the pore to observe the transition of the
frequency spectrum as the effect of water molecule concentration. The
result of our simulation demonstrates that water content in the nanopore
modifies significantly the spectral profile of porous silica system. The
deviation from pure silica vibration spectrum is a direct consequence
from the interaction between silicate based material with the fluid.

Keywords: Molecular dynamics · Rock minerals · Density of states

1 Introduction

The data of mechanical properties of rock material is a key factor in the study on
the physics of solid earth. Up to now, in the geophysical researches the properties are usually derived from the macroscopic physical quantities, obtained from
laboratory or on-site measurements. Atomic scale computational analysis of rock
mechanics is a new approach to understand these properties from a fundamental
level. Theoretically, the physical quantities are collective contributions from a
huge number of infinitesimal partitions of rock material. Each individual partition might have very distinctive characteristics. In measurements based on the
mechanical properties of rock, such as seismology, the quality of data interpretation depends very strongly on the understanding of vibration response of the
material due to mechanical disturbance. The propagation of mechanical wave
through a medium determines the shape of the wave signals received by sensors

PTUPT Project, Ministry of Research Technology and Higher Education, Republic of
Indonesia. Contract No. 2893/UN6.D/LT/2019.

located somewhere on the earth surface. Regarding the aim of such measurements the microscopic information is hidden, or removed from the signals, hence inaccessible for further analysis. However, a microscopically detailed information of rock properties may enhance the quality of data interpretation. Corroborating with the atomistic modeling, the information of the mechanical properties can be made more accurate.

With the current measurement techniques, we have been able to measure the elastic properties in a great detail. Nevertheless, atomistic observation of the vibration mechanism in a rock material is crucial to give more insight which is inaccessible to conventional experiments. Using a good interatomic interaction model, the mechanical properties can be directly obtained, e.g. from the theory of elasticity. This atomic interaction defines fundamentally the physical properties of the building blocks of a medium that can be verified by experimental methods at a larger scale.

In geophysical surveys, the vibration analysis relies on low frequency response of an observed medium. A signal filtering mechanism is commonly applied to remove unintended high frequencies. The vibration signal analysis is very successful to reconstruct the subsurface and the inner-structure of the earth, based only on the reflection and refraction of mechanical wave propagation. To obtain an accurate profile of a considered propagation medium, i.e. at a seismic measurement area, an inversion procedure is necessary. The algorithm depends on accurate information of elastic properties of the medium. Usually, these properties are obtained using borehole analysis of the forming rocks on the measurement site.

Information about physical characteristics of rock materials, contained in high frequency vibration spectrum, can be evaluated microscopically. The methods to evaluate high frequency vibration of material reveal many details on the physical properties that can bring high level of accuracy to the macroscopic measurement. Hence, it is important to calculate rock material properties, in this case silica, from the atomistic approach. The result shall uncover information on the atomic vibration of a complex structure, e.g. a porous material infiltrated by fluid. The advancement of computer technology and physical modeling techniques enables us to run a sufficient size simulation to obtain a detailed data that is comparable to a related experiment result.

The porosity of rock plays an important role on shaping the behavior of wave propagation in a medium [3]. It is clear that the existence of pores attenuates the amplitude and decreases the effective speed of a traveling wave. The existence of pores give rise to complexity of wave mechanics, hence difficult to handle theoretically especially on the case of inhomogeneous pore distribution. The complexity also increases when the pores are filled with other materials, such as metals, water, or hydrocarbons. Using the molecular dynamics technique it is possible to observe and analyze the mechanism in a very great detail.

In this paper we present molecular dynamics simulation result focused on the effect of water content inside the pore of porous silica. The transition of the correlation function and the vibration frequency spectrum from dry to water saturated medium is discussed. The target under consideration is in amorphous phase, hence the result can be related to non-crystalline silicate minerals. The work is intended to give an atomistic analysis on the physical properties of rock in nanometer size, which will contribute to the understanding of water–silica interaction.

2   Method

We performed a series of classical molecular dynamics simulation of porous SiO_2 material in order to study its vibration characteristics. The target material has one percolated pore, filled with water molecules with varied values of density ranging from $\rho = 0$ to the equilibrium density of water in gas phase, $\rho_0 = 0.0256$ Å^{-3}. The variation of density is required to observe the transition from dry to water saturated porous material. The vibration spectrum is evaluated by calculating the Fourier transform of the velocity auto-correlation function. The trajectories of the atoms in the simulation system is followed up to 500 ps. The chosen simulation time is suitable to analyze vibration frequencies higher than 1 THz. To obtain data at lower frequency range with a good statistical confidence requires longer simulation time. Therefore, we do not analyze frequencies lower than this value.

We apply Reax Force Field (*ReaxFF*) potential to model the interaction between atoms. This force field potential allows bond creation and bond breaking of molecules. On that account, it is very suitable to be used to study the chemical processes occurred at the interface between the silicate material and the fluid, as well as the physical response induced by the dynamics of chemical bonds between the surface of the porous material and water molecules. The force field is more computationally convenient in comparison to *ab-initio* or DFT calculation, with an acceptable accuracy [1,10,14]. Using classical molecular dynamics technique, the integration of the force field allows calculation of interactions involving large number of atoms, using considerably low number of CPU cores.

ReaxFF potential consists of terms to handle the complete picture of molecular motion, led by the interaction between atoms. In brief, the total interaction potential U_{tot} is expressed as,

$$U_{tot} = U_{bond} + U_{vdw} + U_{coul} + U_{over} + U_{angle} + U_{tors} + U_{spec}, \qquad (1)$$

where the indices indicate the type of interaction, i.e. bonding between atoms in a molecule, the long-ranged *van der Waals* interaction between neutral atoms, the long-ranged *Coulombic* interaction between charged atoms, the energy penalty for over coordinated atoms, the angular interaction of three bonded particles, the torsion and the specific energy, respectively. The last term, U_{spec}, is available to accommodate energy compensation that is specific to the system of interest. In this work we use the potential parameter described in [2], which is specially fitted to simulate water-silica interface.

The vibration pattern of the system is observed through the velocity auto-correlation function, which is defined as,

$$C_v(t) = \frac{1}{N} \sum_{k+1}^{N} \langle \boldsymbol{v}_k(t) \cdot \boldsymbol{v}_k(0) \rangle \tag{2}$$

for N number of particles in the system. The angle bracket represents the time average. Since the correlation function is invariant to the time shift, the velocity $\boldsymbol{v}(0)$ is not necessarily taken at the initial time. The vibration frequency density $\Psi(\nu)$ is calculated from the function, such that,

$$\Psi(\nu) = \int_{-\infty}^{\infty} C_v(t) e^{-2\pi i \nu t} dt \tag{3}$$

which is the Fourier transform of the correlation function [8]. The frequency, ν, is measured in Hertz. In order to eliminate fluctuations we apply moving average filtering algorithm to the transformed data. We may write,

$$\Psi(\nu_{j+n_a/2}) = \frac{1}{n_a} \sum_{k=j}^{j+n_a} \Psi(\nu_k) \tag{4}$$

where n_a is the number of averaging points, and the index in ν denotes the discreet index of frequency data. In the frequency spectrum of the porous target we take $n_a = 100$.

Table 1. The geometrical configuration of the target material.

Target	Number of atom	Size (Å^3)
Silica bulk	3960 (Si) 7920 (O)	$54 \times 51 \times 54$
Water vapor	10000 (H) 5000 (O)	$58 \times 58 \times 58$
Porous silica	36347 (Si) 73874 (O)	$148 \times 145 \times 146$

In this work the MD simulation was performed using the publicly available LAMMPS code [9]. The analysis was done using the Open Visualization Tool (OVITO), both the graphical user interface and the Python scripting tool [13]. To fill water molecules inside the pore we used PACKMOL program [7]. The system configurations and the target size used in the series of simulations are shown in Table 1. For the water filling we put a desired fraction of ρ_0 inside the pore, which occupies 50% of the porous silica volume.

2.1 Target Material Preparation

The Preparation of a-SiO$_2$ Target Material. We follow the recipes described in [5] in order to obtain amorphous silicate phase. The procedure consists of heating and cooling sequence of a target crystal. We start the process from an α-quartz

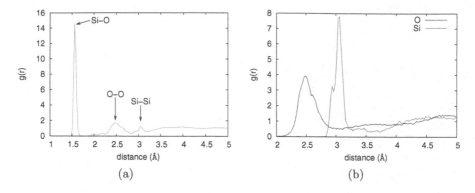

Fig. 1. (a) The total and (b) the partial radial distribution function for the annealed-quenched target material.

structure followed by annealing to a temperature of 8000 K. This temperature is required in order to break the crystal configuration while keeping the correct local molecular structure (SiO_4, tetrahedral). After the annealing process, the hot crystal is brought to 300 K by quenching mechanism. The whole procedure is done in NPT ensemble, where the number of atoms, pressure and temperature of the system are kept constant. During this process the temperature is monitored and the simulation is stopped when the desired temperature is reached and the fluctuation is stable. In the final target material there is a fraction of under-coordinated atoms of about 25%. Figure 1 presents the partial and total radial distribution function (RDF) of the amorphous silicate. The peaks show the correct bond distance of Si-O, O-O, and Si-Si atoms, which is 1.6 Å, 2.5 Å, and 3.1 Å, respectively. This indicates that the silicate material is in the correct local structure.

The radial distribution function of amorphous SiO_2 has been evaluated by Hoang in [4], and the data agrees very well with our result. The first peak of the RDF plot identifies the distance between silicon and oxygen atom inside one SiO_4 tetrahedron. The second and the third peaks identify bonds between the next nearest neighbors with a same type. In the partial RDF (c.f. Fig. 1b) these bond distances are more clearly shown.

Figure 2a shows an exemplary snapshot of local configuration of an amorphous silica. The atom coordination number is calculated by evaluating the type and distance of neighboring atoms in a shell with radius of Si-O distance, shown as the first peak in the RDF plot. The iteration is run through all Si atoms. Atoms having more than 4 oxygen neighbors are marked as over-coordinated, and those having less than 4 as under-coordinated ones. In the final target we obtain 25% under-coordinated and no over-coordinated Si atoms, from the total number of atoms.

In contrast to single element materials, amorphous SiO_2 maintains the local SiO_4 tetrahedral structure. The amorphization happens by tilting the angle at a shared oxygen atom, i.e. the angle created by Si-O-Si bonds. In a quartz structure the angle is 144°. The annealing process disperse the angular distribution ranged

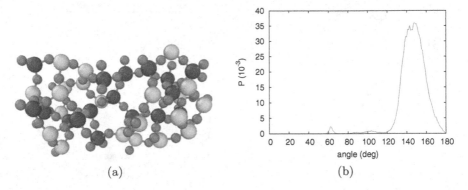

Fig. 2. (a) A snapshot of local structure in an amorphous silica. Red balls represent Si atoms as a center of tetrahedra with 4 coordination number. Yellow balls represent under-coordinated Si atoms. Blue balls are oxygen atoms (b) The angular distribution of Si-O-Si bond, representing the tilting between two SiO_4 tetrahedra. (Color figure online)

from around $120°$ to $180°$. Figure 2b shows the plot of angular distributions of a quenched target at $300\,K$ after annealing. Some part of the bonds create a small bump at around $60°$, which demonstrate the probability of triangular Si-O-Si structure, where one oxygen atom is shared between two bonded Si atoms.

The Preparation of Porous Material. The porous material is generated by sequential annealing and quenching procedure using an artificial Lennard-Jones material [11]. Firstly, we create an FCC crystal and remove a number of random particles according to the desired porosity (here 50%). An MD simulation is performed using NPT ensemble to heat the system until reaching the melting temperature. We let the simulation run until the fluctuation stays less than 10% around the desired temperature. We apply periodic boundary condition to the system. Secondly, the system is quenched to zero temperature to allow the porous meta-material relax to a stable phase. We use potential parameters with an unphysically high binding energy in order to allow the particle to stick together. This procedure is to assure elongated filaments and the pore to percolate through the sides of the porous model. After the process, we take only the largest cluster in the quenched target using a cluster detection algorithm [12]. Finally, all small clusters and left-over atoms are removed. The position of particles in this porous meta-material are mapped into a 3D porous mask. In this case we intuitively choose an FCC structure for the mask and then used as a negative template to make porous material model. The template can be scaled further to match the desired size and mapped to the actual target material. The creation of target material can be easily done by overlaying the template and the target together, and then all the overlapping particles are removed. Although, a rigorous mathematical method to create a porous model is available [6], we apply this method to mimic random porous structure which is a close approximation to a naturally formed porous rocks.

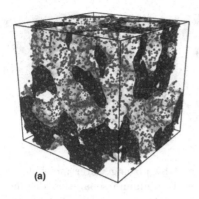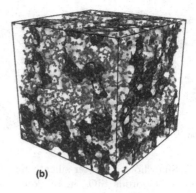

Fig. 3. Snapshots of the created porous material (a) before relaxation, (b) after relaxation. Blue circles are oxygen atoms and red circles represent the silicon atoms. White area indicates the pore surface. (Color figure online)

We minimize the number of under-coordinated silicon at the surface in order to maintain the tetrahedral SiO_4 structures. This leaves dangling bonds of oxygen atoms which naturally capture hydrogen atoms to form silanol functional group [10]. Using the previously created porous template, we remove only the overlapping Si atoms in the target to make sure that oxygen atoms belong to a tetrahedron at the surface are still intact. This leaves free oxygen atoms inside the pores. Since the distance of leftover oxygen is easily identified, we can select atoms having more than Si-O bond distance and remove from the target. After this final removal, we relax the system using NPT ensemble to room temperature. The relaxation step can not avoid evaporation of some oxygen atoms, due to the abrupt removal that may leave large amount of kinetic energy at the surface. Figure 3 shows snapshots of the porous target material before and after annealing. The pore size changes due to thermal coarsening during the relaxation process.

We used PACKMOL program to put water molecules in random position and orientation inside the pores. The number of molecules is calculated according to the desired density of water inside the pore. We define the tolerance distance of $r_{tol} = 2$ Å to avoid assigning the molecules at the interstitial sites of the SiO_2 bulk. After deposition of some amount of water molecules inside the pores we obtain a water-silicate matrix. Figure 4 shows snapshots of the matrix with different density of water. The snapshots are taken after quenching and relaxation procedure at room temperature. The simulation is executed up to 500 ps simulation time, while keeping the temperature constant at 300 K with zero pressure using NPT ensemble.

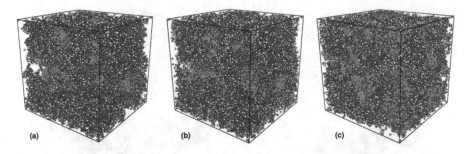

Fig. 4. Snapshots of water-silicate matrix created by placing water molecule inside the pores of porous SiO_2 in random position with minimum separation distance of $r_{tol} = 2$ Å, (a) for $\rho = 0.25\rho_0$, (b) $\rho = 0.5\rho_0$, and (c) $\rho = \rho_0$. Blue balls represent oxygen, yellow silicon, and red hydrogen atoms. (Color figure online)

3   Result and Discussion

3.1   The Vibration Density of the Pure Material

The vibration characteristic of bulk SiO_2 material is used as a reference to see the transition and the effect of water filling of the porous material. Figure 5a shows the plot of vibration density of states calculated through the Fourier transform of the velocity auto-correlation function. We focused at high frequency region of the spectrum for $\nu > 10$ THz. The peak height in the spectrum of the amorphous phase decreases due to lost of regularity in atomic positions which alter the vibration behavior. However, the locations of dominant frequencies stay the same. At frequencies $\nu > 100$ THz new peaks appear, indicating that the amorphousity induces vibration at a higher frequency. The frequencies lay beyond the cut-off frequency of the crystalline Si around $\nu = 25$ THz [8]. High probably, this vibration is the effect of bond defects, such as dangling bonds of under-coordinated atoms.

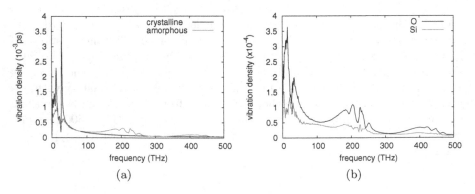

Fig. 5. (a) The plot of vibration density of crystalline and amorphous silicate measured using the velocity auto-correlation method, normalized to the area below the curve. (b) Partial vibration density of amorphous SiO_2. This plot is not normalized.

Figure 5b shows the partial vibration spectrum of the amorphous target. In this plot, we do not normalize the data to make it easier to compare the amplitudes. From the plot we can conclude that oxygen atoms dominate the vibration of the material at all range of frequency. A dip at $\nu \cong 10$ THz is characteristics to SiO_2 material, which appears clearly in the crystalline density of states. Silicon atoms seem to mimic the vibration of oxygen atoms, shown by small bumps at the same location at high frequency zone. These bumps indicate coupled atomic vibration of two atom species. It is clear that the smaller atoms are able to move with larger excursion from its equilibrium bond length. This induces higher magnitude in the vibration spectrum.

Table 2. The diffusion coefficient (D) and the correlation time (t_{corr}) of the porous material filled with water of density ρ.

ρ/ρ_0	D (Å^2/fs)	t_{corr} (fs)
0	0.002	7.96
0.05	0.100	12.78
0.25	0.080	15.66
0.50	0.076	17.88
1.00	0.047	20.55

3.2 The Vibration Density of Water-Silica Matrix

Figure 6 shows the plot of velocity auto-correlation function C_v, of the porous material with water filling densities from $\rho = 0$ to $\rho = \rho_0$. From the correlation curves, we can deduce the three dimensional diffusion coefficient D by taking the time integral of C_v as,

$$D = \frac{1}{3} \int_0^\infty C_v(t)dt \qquad (5)$$

Table 2 shows the diffusion coefficients calculated by Eq. 5. The value increases immediately at the lowest density of water, $\rho = 0.05\rho_0$, and then decreases with the decreasing value of the density. The large difference between the diffusion coefficient of the empty pore and the water-filled pore indicates that the diffusion process is dominated by water molecules.

In order to measure the characteristic time of atomic motion related to the initial state, we approximate the correlation time t_{corr} by fitting the data to a decay function,

$$A = A_0 \cdot e^{-t/t_{corr}} \qquad (6)$$

where A denotes the amplitude. We observe that the correlation time increases systematically with the increment of the water density. We can define a time

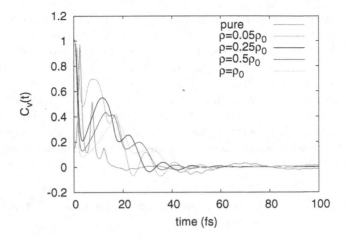

Fig. 6. The velocity autocorrelation function of the SiO_2 porous material filled with water. The "pure" symbol represents the function of a dry material.

scale when the motion of molecules in the system is not correlated to their initial velocity as $t \geq 3 \cdot t_{corr}$.

For a porous SiO_2 material, we observe a same vibration profile compared to the bulk amorphous material. Note that the porous material is also in amorphous phase. The plot of both porous and bulk target is shown in Fig. 7a. The similarity demonstrates that the porous structure does not modify the vibration characteristic of the material, especially for the high frequency range. A small fluctuation occurs at $\nu \cong 200$ THz may be a size effect, since in the simulation of bulk SiO_2 we use much smaller target size. At low frequency region the two curves are slightly different, indicating that the porosity influence to the vibration behavior of the amorphous silica may also lay at low frequency range. However, the analysis of low frequency spectrum is beyond the subject of this

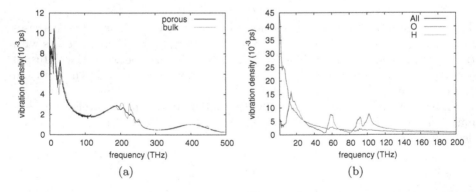

Fig. 7. (a) The vibration density of pure SiO_2 material. (b) the total and partial vibration density of pure H_2O.

work. Figure 7b is the vibration density of water vapor. The partial density profile shows that the characteristic peaks are dominated by the H atoms, since the oxygen atoms are less mobile.

Figure 8 shows a complete set of vibration densities of the porous material filled with water. The plots are normalized to the area below the curves. The data shows that the profile of vibration spectrum of the dry material is suppressed by the existence of water. This finding is obvious regarding the higher mobility of water molecules in comparison to silica in the matrix. This is also indicated from the higher value of diffusion coefficient of water in comparison to the SiO_2 matrix. From the figure we can determine at least two dominant frequencies, a broader peak at $\nu = 4$ THz and a sharp peak at $\nu = 50$ THz. The peaks remain stable with small influence from the density variation. It is evidence that they are correlated with the vibration properties of water. However, these frequency peaks do not appear in the pure vibration density of silica and water (c.f. Fig. 7), hence they are characteristics to the SiO_2–H_2O compound system.

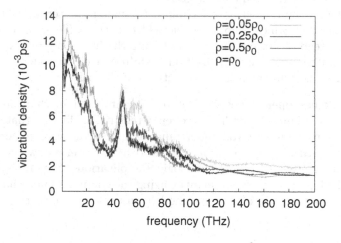

Fig. 8. The vibration density profile of SiO_2 porous material filled by water vapor with variation on the density, $\rho = 0$ to $\rho = \rho_0$. The plots are normalized by the area below the curve.

There is a build up of new peaks, at around $\nu = 30$ THz, $\nu = 60$ THz, and $\nu = 90$ THz, that coincides perfectly with water frequency profile. The peaks are becoming more apparent along with the increment of the water density. Hence, we may conclude that these peaks appear due to the vibration of water molecules. Notably, the characteristic peaks of a pure SiO_2 material are heavily suppressed. The saturation of a porous silica material by water induces a new vibration profile, which is different from the base material.

4   Conclusion

We performed a set of MD simulations to investigate the effect of water saturation on a porous silica material, in a form of water-silica matrix. The data shows a significant deviation on the vibration characteristics of the matrix, induced by water molecules inside the pore of amorphous silica. From the result of our simulation we can conclude that,

1. The existence of water molecules suppresses the vibration characteristics of the base material. The frequency peaks of the molecules appears gradually with the increment of water density.
2. The vibration spectrum is dominated by the low mass atoms. On SiO_2, oxygen bond leads to vibration at high frequency region (around $\nu = 200$ and 400 THz) beyond the cut-off frequency of a pure silicon. On water the spectrum profile is completely governed by the O-H bond. It is shown by the partial vibration density of states that the bond with light atoms dominates the amplitude of the frequency spectrum.
3. The profile of the frequency spectrum of the matrix material (SiO_2-H_2O) is newly generated, which is not constructed from the profile of its base material (SiO_2). Instead of attenuating or amplifying the frequency peaks, in a multi element matrix material the vibration is shifted or created at new frequency value, which is different from the spectrum of the base material.

This work has opened new insight to the vibration mechanism of multi-element compound material at high frequency region. The finding that is revealed from the data may improve the understanding on the vibration of complex systems such as natural rocks or laboratory fabricated materials, which may have impact on the enhancement of technological applications. We emphasized that such a detailed analysis is available by utilizing the computation technology that leads to the improvement of material models and calculation methods.

References

1. Chenoweth, K., van Duin, A.C.T., Goddard III, W.A.: ReaxFF reactive force field for molecular dynamics simulations of hydrocarbon oxidation. J. Chem. Phys. A **112**, 1040 (2008)
2. Fogarty, J.C., Aktulga, H.M., Grama, A.Y., Van Duin, A.C., Pandit, S.A.: A reactive molecular dynamics simulation of the silica-water interface. J. Chem. Phys. **132**(17), 174704 (2010). https://doi.org/10.1063/1.3407433
3. Gunkelmann, N., Bringa, E.M., Rosandi, Y.: Molecular dynamics simulations of aluminum foams under tension: influence of oxidation. J. Phys. Chem. C **122**(45), 26243–26250 (2018). https://doi.org/10.1021/acs.jpcc.8b07146
4. Hoang, V.V.: Molecular dynamics simulation of amorphous SiO2 nanoparticles. J. Phys. Chem. B **111**, 12649 (2007)
5. Huff, N.T., Demiralp, E., Çagin, T., Goddard, W.A.: Factors affecting molecular dynamics simulated vitreous silica structures. J. Non-Cryst. Solids **253**(1–3), 133–142 (1999). https://doi.org/10.1016/S0022-3093(99)00349-X

6. Liu, C., Branicio, P.S.: Efficient generation of non-cubic stochastic periodic bicontinuous nanoporous structures. Comput. Mater. Sci. **169**, 109101 (2019). https://doi.org/10.1016/j.commatsci.2019.109101

7. Martinez, J.M., Martinez, L.: Packing optimization for automated generation of complex system's initial configurations for molecular dynamics and docking. J. Comput. Chem. **24**, 819–825 (2003)

8. Meyer, R., Comtesse, D.: Vibrational density of states of silicon nanoparticles. Phys. Rev. B - Condens. Matter Mater. Phys. **83**, 014301 (2011). https://doi.org/10.1103/PhysRevB.83.014301

9. Plimpton, S.: Fast parallel algorithms for short-range molecular dynamics. J. Comput. Phys. **117**(1), 1–19 (1995). https://doi.org/10.1006/jcph.1995.1039

10. Rimsza, J.M., Yeon, J., van Duin, A.C.T., Du, J.: Water interactions with nanoporous silica: comparison of ReaxFF and ab initio based molecular dynamics simulations. J. Phys. Chem. C **120**(43), 24803–24816 (2016). https://doi.org/10.1021/acs.jpcc.6b07939

11. Rosandi, Y., Wijatmoko, B.: Generation of nanoporous model using sequential annealing and largest cluster selection method. IOP Conf. Ser. J. Phys. **1080**, 012027 (2018)

12. Stoddard, S.D.: Identifying clusters in computer experiments on systems of particles. J. Comput. Phys. **27**(2), 291–293 (1978). https://doi.org/10.1016/0021-9991(78)90011-6

13. Stukowski, A.: Visualization and analysis of atomistic simulation data with OVITO-the open visualization tool. Model. Simul. Mater. Sci. Eng. **18**(1), 015012 (2009). https://doi.org/10.1088/0965-0393/18/1/015012

14. Yeon, J., Van Duin, A.C.T.: ReaxFF molecular dynamics simulations of hydroxylation kinetics for amorphous and nano-silica structure, and its relations with atomic strain energy. J. Phys. Chem. C **120**, 305–317 (2016). https://doi.org/10.1021/acs.jpcc.5b09784

Numerical Study of Dispersive Mass Transport in Homogeneous and Heterogeneous Porous Media

Hector Rusinque$^{(\boxtimes)}$, Eugenia Barthelmie, and Gunther Brenner

Institute of Applied Mechanics, Clausthal University of Technology,
38678 Clausthal-Zellerfeld, Germany
hector.rusinque@tu-clausthal.de,
http://www.itm.tu-clausthal.de

Abstract. A modular simulation approach is used to compute the flow of a fluid and the mass transport of tracers in the void space of computer-generated porous packings. Effective transport properties such as the diffusive tortuosity and the dispersion tensor are determined. First, we present and compare two different approaches to model mass transport in homogeneous porous media. Subsequently, heterogeneous porous media are considered, where we investigate the effect of walls on the structure of confined random sphere packings and how it affects the mass transport properties of a sphere packing. In addition, the hydraulic tortuosity is computed and its performance as a descriptor of porous media is compared with that of the diffusive tortuosity.

Keywords: Porous media · Mass transport · Brownian dynamics · Tortuosity · Dispersion coefficient

1 Introduction

A rigorous characterization of porous media is important for the determination of parameters needed in mathematical models that can describe e.g. the spreading of a contaminant in an aquifer, the successful delivery of a drug in a desired tissue or the efficiency of a separation process as in column chromatography. Such a characterization is usually complex, since mass transport properties correlate strongly with the intricate pore structure of the porous medium. Usually, upscaled models and correlations failed to capture the anisotropic nature of heterogeneous porous media [1], several descriptors have been proposed such as the hydraulic tortuosity in the Kozeny-Carman equation [1–5]. There are several models capable of describing transport phenomena in porous media such as the method of volume averaging (MVA) [6–8], Brownian dynamics (BD) [9–12], homogenization [13], and the thermodynamically constrained averaging theory [14]. In this study, we used the MVA and BD approaches in order to describe the motion of molecules diffusing in the void space of homogeneous and heterogeneous porous materials. After solving the differential equations involved in each

© Springer Nature Switzerland AG 2020
N. Gunkelmann and M. Baum (Eds.): SimScience 2019, CCIS 1199, pp. 104–121, 2020.
https://doi.org/10.1007/978-3-030-45718-1_7

approach, effective transport parameters and descriptors of the porous structure
are determined from the obtained information of the system as will be explained
in detail in the modeling section of this work.

Further, both methods are described and compared considering transport in
homogeneous porous media. Finally, wall effects on mass transport are studied.
For this purpose, confined sphere packings were computer-generated and their
mass transport properties were obtained using the BD approach.

2   Modeling and Numerical Approach

We commence by describing the general system with domain $\Omega \subset \mathbb{R}^3$ or in some
simulation cases $\Omega \subset \mathbb{R}^2$. It is considered a rigid porous medium (a solid phase
Ω_s) completely filled with a fluid phase with domain Ω_f. The fluid flows through
the porous medium carrying a solute (passive point-wise species), which in turn
moves via diffusion. The solid phase is assumed impermeable to mass transport.

We combined two simulation steps in both computational approaches (MVA
and BD). In the first step, the velocity field is computed. In the second simulation
step, we simulate the mass transport using the velocity field obtained in the first
step, as input.

2.1   Computation of the Velocity Field

We obtained the velocity fields from the numerical solution of the Stokes equation [15,16] or the Boltzmann equation, using the Lattice Boltzman method
(LBM) [17,18]. As all the porous structures considered are periodic in at least
two directions, we used periodic boundary conditions for the pressure and velocity fields on the respective boundaries. The no-slip condition for viscous flow
was applied on the walls and solid-fluid interface. A constant external force was
used as driving force inducing the flow. For the LBM, we chose a force so that
the Reynolds number is kept low enough, i.e., Re $\ll 1$, to assure creeping flow.
Advective inertial forces are not considered in the Stokes equations per default,
resulting in the following equations

$$\sum_{i=1}^{3} \frac{\partial u_i}{\partial x_i} = 0 , \tag{1}$$

$$\frac{\partial p}{\partial x_j} - \sum_{i=1}^{3} \left(\mu \frac{\partial^2 u_j}{\partial x_i^2} \right) = f_j \quad \text{for } j = 1, 2, 3. \tag{2}$$

We used the open-source computing platform FEniCS [19,20] for solving the
Stokes equation and an in-house solver for the LBM [21–23].

2.2   Computation of the Mass Transport

The two different approaches presented in this part can be used to simulate mass transport in porous media. The first approach, the MVA is based on the solution of the volume-averaged convection-diffusion equation [6–8]. The second approach uses Brownian dynamics and is based on the solution of the Langevin equation for passive tracers [9–12].

Dispersion via the Method of Volume Averaging. Here, we considered the fluid domain Ω_f with boundary $\partial\Omega_f = \partial\Omega_{fe} \cup \partial\Omega_{fs}$. $\partial\Omega_{fe}$ stands for the entrances and exits of the fluid phase, whereas $\partial\Omega_{fs}$ represents the fluid-solid interface. For the fluid domain, the microscale convection-diffusion equation for mass transport of a scalar (i.e., point-wise) species A is given by

$$\frac{\partial C_A}{\partial t} + \sum_{i=1}^{3} u_i \frac{\partial C_A}{\partial x_i} = \sum_{i=1}^{3} D_\infty \frac{\partial^2 C_A}{\partial x_i^2}, \text{ in } \Omega_f \tag{3}$$

$$-\sum_{i=1}^{3} n_i D_\infty \frac{\partial C_A}{\partial x_i} = 0, \text{ at } \partial\Omega_{fs} \tag{4}$$

$$C_A = C_e(x_i, t), \text{ on } \partial\Omega_{fe} \tag{5}$$

$$C_A = C_0(x_i), \text{ when } t = 0 \tag{6}$$

In the above equations, C_A is the molar concentration of species A, D_∞ the unbounded diffusion coefficient (or mixture diffusion coefficient in case of a mixture), u_i the velocity field. The subindex i represents the elements of the velocity vector in Cartesian coordinates, e.g. $u_i = (u_x, u_y, u_z)$.

After applying the MVA to the microscopic convection-diffusion equation [6], we obtain the following upscaled equation for the case of a homogeneous porous medium [6,8]

$$\frac{\partial \langle C_A \rangle^f}{\partial t} + \sum_{i=1}^{3} \langle u_i \rangle^f \frac{\partial \langle C_A \rangle^f}{\partial x_i} = \sum_{i=1}^{3} \frac{\partial}{\partial x_i} \left(D_{ij}^* \frac{\partial \langle C_A \rangle^f}{\partial x_j} \right), \tag{7}$$

with $j = 1, 2, 3$, and dispersion tensor D_{ij}^*

$$D_{ij}^* = D_\infty \left(\delta_{ij} + \frac{1}{\Omega_f} \int_{\partial\Omega_{fs}} n_i b_j^* \, dA \right) - \langle \tilde{u}_i b_j^* \rangle. \tag{8}$$

where $\tilde{u}$ stands for the spatial deviation of the velocity field, b_i^* is the associated closure variable and δ_{ij} the identity matrix.

Closure Variable. The closure variable solves the following boundary-value problem in a representative periodic cell [7,8].

$$\sum_{i=1}^{3} u_i \frac{\partial b_j^*}{\partial x_i} = \sum_{i=1}^{3} \left(D_\infty \frac{\partial^2 b_j^*}{\partial x_i^2} \right) - \tilde{u}_j, \quad \text{in } \Omega_f \tag{9}$$

with boundary and periodicity conditions for the fluid phase

$$-\sum_{i=1}^{3} n_i D_\infty \frac{\partial b_j^*}{\partial x_i} = n_j D_\infty, \quad \text{at } \partial\Omega_{fs} \tag{10}$$

$$b_j^*(x_k + l^k) = b_j^*(x_k), \quad k = 1, 2, 3, \tag{11}$$

and following constraint, which is needed for numerical stability and consistency

$$\langle b_j^* \rangle^f = 0. \tag{12}$$

This closure problem (Eqs. 9–12) can be interpreted as a transport equation describing the convective and diffusive transport of the vectorial entity b_i^* with two sources. On the one hand, we have a convective volume source being the velocity deviation field $\tilde{u}$. This source can be negative or positive depending on the deviation of the local velocity with respect to the average velocity. In fact, its average over the volume of the fluid phase Ω_f is zero [6]. On the other hand, we have a diffusive surface source $n_i D_\infty$ whose sign depends on the orientation of the normal vector n_i. Similarly, the average value of the surface source over $\partial\Omega_{fs}$ is zero [6].

The effective diffusivity D_{ij}^{eff} and diffusive tortuosity τ_{ij} are defined using the MVA with the diffusion equation, i.e., Eqs. 3–6 for the case of no convection $u_i = 0$, as follows (see averaging procedure by Whitaker in [6])

$$D_{ij}^{\text{eff}} = \frac{D_\infty}{\tau_{ij}} = D_\infty \left(\delta_{ij} + \frac{1}{\Omega_f} \int_{\partial\Omega_{fs}} n_i b_j \, dA \right), \tag{13}$$

where the closure variable b_i is obtained from the boundary-value problem

$$0 = \sum_{i=1}^{3} \left(D_\infty \frac{\partial^2 b_j}{\partial x_i^2} \right), \quad \text{in } \Omega_f \tag{14}$$

$$-\sum_{i=1}^{3} n_i D_\infty \frac{\partial b_j}{\partial x_i} = n_j D_\infty, \quad \text{at } \partial\Omega_{fs} \tag{15}$$

$$b_i(x_k + l^k) = b_i(x_k), \quad k = 1, 2, 3, \quad \text{periodicity.} \tag{16}$$

Note that this implies that the diffusive tortuosity only depends on the geometry of the porous structure. Further, from the definition of the dispersion tensor

(Eq. 8), one can conclude that dispersion not only depends on the geometry of the porous medium but also on the spatial deviation of the velocity. In fact, for high Péclet numbers, the axial component of the dispersion tensor (here D_{xx}) is strongly dominated by the velocity deviation, as shown in the results section of this paper. Notice that dispersive transport is not directly the superposition of convective and diffusive transport but rather the combined dispersive effect of the spatial deviation of the velocity and diffusion. A simple scenario to illustrate this phenomenon is a (non-viscous) plug flow where we will not observe any enhancement of the dispersive effect by increasing the Péclet number as the spatial deviation of the velocity field is zero. In contrast, in case of viscous pipe flow the dispersive effect of the diffusive transport is combined to that of the deviation of the velocity with respect to the mean flow velocity. The latter case is known as Taylor dispersion [24].

As a descriptor for the hydrodynamic dispersion we have to use one that evaluates the velocity field. The hydraulic tortuosity τ_h was used to describe the anisotropy of porous structures correlating it with e.g. the permeability tensor [1–5]. We have computed the hydraulic tortuosity according to the following equation [25,26].

$$\tau_{h_x} = \frac{\langle \sqrt{u_x^2 + u_y^2 + u_z^2} \rangle}{\langle |u_x| \rangle} \qquad (17)$$

In order to solve the closure problems for b^* and b, we extended the FEM-based FEniCS solver [19,20]. Thus, after computing the velocity and velocity deviation fields (u_j and $\tilde{u}_j$), the results are given as input to the b^*-solver.

Dispersion via Brownian Dynamics. We define the equation of motion of a tracer with position x_i and velocity v_i, which moves in the fluid phase through the porous network. This is given by the Langevin equation

$$m\frac{\partial v_i}{\partial t} = \sqrt{2\gamma_s k_B T} W_i(t) - \gamma_s (v_i - u_i) - \frac{\partial U}{\partial x_i}, \quad i = 1,2,3, \qquad (18)$$

where m is the mass of the tracer, γ_s is the drag coefficient, and $U(x_i)$ the particle interaction potential whose negative gradient $-\partial U / \partial x_i$ represents the force induced by the potential. T and k_B stand for the temperature and Boltzmann constant, respectively.

The first term on the right side of the equation represents the thermal-driving force whose stochastic behavior is modeled by Gaussian noise $W_i(t)$, a normally distributed random number with zero mean $\mu = 0$ and variance $\sigma^2 = 1$ [27,28]. The thermal force accounts for the effect of the solvent on the tracer.

The second term on the right represents the drag force, which is proportional to the relative velocity of the particle with respect to the bulk flow ($v_i - u_i$).

Here, inertial effects can be neglected since the time scale considered spans from the inertial ballistic regime all the way to the diffusive regime [29,30].

This is supported by the fact that the average motion of a Brownian particle over time and the average over the particle ensemble are analogous according to the ergodic hypothesis [31,32], because all effective quantities computed are an average over the total number of tracers. Furthermore, no interaction potential U is considered since we want to model transport of passive scalars as in the case of the volume-averaged convection-diffusion equation discussed in the previous section. Applying these simplifications and the term $v_i = \frac{dx_i}{dt}$ the equation of motion can be rewritten to

$$0 = \sqrt{2\gamma_s\, k_{\mathrm{B}}T} W_i(t) - \gamma_s \left(\frac{dx_i}{dt} - u_i \right), \tag{19}$$

with initial conditions as follows

$$x_i = X_{0,i}, \quad \text{when } t = 0, \tag{20}$$

$$v_i = V_{0,i}, \quad \text{when } t = 0, \tag{21}$$

where $X_{0,i}$ and $V_{0,i}$ are vectors containing the initial positions and velocities, respectively. Note that in Brownian dynamics (BD) the temperature can be controlled as with a thermostat, thus approximating the canonical ensemble. The above initial-value problem was numerically integrated by applying the leapfrog method [33] using an in-house C++ code parallelized with MPI and OpenMP. The velocity field of the bulk flow u_i was taken from the solution of the LBM.

By tracking the position of the particles with the Lagrangian method, one can compute the mean square displacement $\langle \Delta x_{ij}^2(t) \rangle$ of the particle ensemble

$$\langle \Delta x_{ij}^2(t) \rangle = \frac{1}{N} \sum_{n=1}^{N} (x_i^n(t) - \langle x_i(t) \rangle)(x_j^n(t) - \langle x_j(t) \rangle), \tag{22}$$

where $\langle ... \rangle$ indicates averaging over the tracer ensemble and N is the number of tracer particles. From its time derivative, the elements of the dispersion tensor D_{ij} can be calculated [34]

$$D_{ij} = \frac{1}{2} \frac{d}{dt} \langle \Delta x_{ij}^2(t) \rangle. \tag{23}$$

Notice that the off-diagonal elements of the dispersion tensor (D_{ij} for $i \neq j$) vanish when the axial and radial axes of the diffusion ellipsoid coincide with the frame of reference of the diagonal terms of the tensor [35]. In the present work this is approximately the case, since the axial axis of the diffusion ellipsoid is predetermined by the selected main direction of the bulk flow.

In this study, only the transport of passive scalar tracers is considered. Deviations from scalar transport can be observed, e.g. when the pore size comes too close to the tracer particle size [12,32,36], or the tracer shape diverges strongly from a sphere, as well as when pairwise interactions between the diffusing particles are strong and cannot be neglected [37–39].

3   Results and Discussion

3.1   Mass Transport in Homogeneous Porous Media: Comparison of the MVA and BD Approaches

The flow through an arrangement of in-line cylinders is considered (see Fig. 1). The porosity was varied by increasing the cylinder diameter.

As required by the definition of the closure variables b and b^* in the MVA, the system consists of a representative periodic unit cell of the homogeneous porous medium considered. The relative values of the radial and axial dispersion coefficients with respect to the free (unbounded) diffusion coefficient (D_{xx}/D_∞ and D_{yy}/D_∞, respectively) were determined with increasing Péclet number (ratio of convection to diffusion).

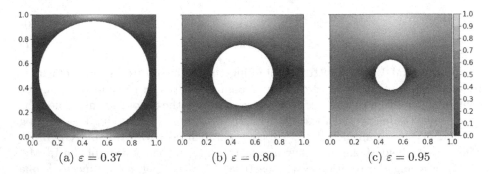

(a) $\varepsilon = 0.37$ (b) $\varepsilon = 0.80$ (c) $\varepsilon = 0.95$

Fig. 1. Periodic unit cell of the in-line cylinder arrangements with a graphic representation of the velocity field for three porosity values. The distance in the x and y axes, and velocities were normalized by the maximal magnitude of the distance and velocity field, respectively.

Dispersion Regimes

Low Péclet numbers Pe < 1 (i.e., low bulk-flow velocities) this dispersion regime is controlled by diffusion, as convection is weak here. Thus, in this range, the value of the dispersion coefficients is determined solely by the geometry of the porous structure, making the diffusive tortuosity the descriptor par excellence for this region. In fact, the relative dispersion coefficients, D_{xx}/D_∞ and D_{yy}/D_∞, converge to the reciprocal value of the diffusive tortuosity (τ_{xx}^{-1} and τ_{yy}^{-1}) as the Péclet number (Pe) goes to zero. This can be clearly seen in Fig. 3 in comparison to the values of the diffusive tortuosity shown in Table 1. The diagonal components of the tensor of diffusive tortuosity τ_{ii} assume the same value since the porous medium is isotropic, i.e., $\tau_{xx} = \tau_{yy}$.

The diffusive tortuosity is related to the degree of diffusive paths obstructed by obstacles (the solid boundaries of the pores). For this reason, its reciprocal value is also known in literature as obstruction factor. In general, the diffusive tortuosity has a hindering effect on dispersion, i.e., the higher the tortuosity the lower the value of the dispersion coefficient in this diffusive regime (see Eq. 13).

High Péclet numbers Pe > 1 (i.e., high bulk-flow velocities) Here the axial dispersion is dominated by the spatial deviations of the velocity field. Two main dispersion phenomena can be distinguished that are caused by these spatial deviations. The first is due to velocity gradients between the different stream paths along the porous medium (e.g. trans-column, trans-channel, inter-channel, and Taylor dispersion [24]). The second effect is induced by the splitting of the flow paths and is known as mechanical dispersion. These phenomena have an enhancing effect on the dispersion coefficients.

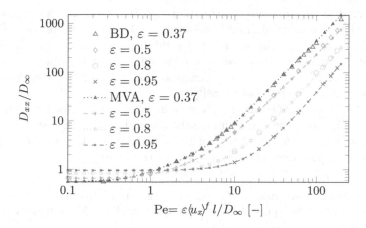

Fig. 2. Axial dispersion coefficients as a function of the Péclet number in the in-line cylinder arrangements considered.

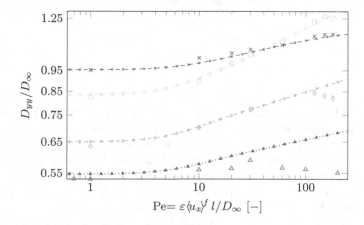

Fig. 3. Radial dispersion coefficients as a function of the Péclet number in the in-line cylinder arrangements considered (see legend in Fig. 2).

Since the velocity field is the determining parameter in this hydrodynamic regime, we will try to establish a connection between the dispersion coefficient and the hydraulic tortuosity in the next section.

Diffusive vs. Hydraulic Tortuosity. The diffusive and hydraulic tortuosity values are shown in Table 1. It is observed that an increase in porosity leads to a higher diffusive tortuosity, which inversely correlates with the dispersion coefficient in the diffusive regime (Pe < 1) as explained above. Regarding the hydrodynamic regime, at Pe > 1, an increase of the hydraulic tortuosity is likewise observed as the porosity increases. Here, the axial dispersion coefficient increases with increasing hydraulic tortuosity and increasing porosity. This correlation is so strong that the order of the axial dispersion coefficients by porosity reverses at about Pe = 1. However, the sensitivity of the hydraulic tortuosity is very small as can be seen from the fact that the hydraulic tortuosity decreases only slightly by the increase of porosity.

The radial dispersion coefficient shows a more complex behavior, as the radial velocities of the bulk flow are significantly lower than the axial velocities. Consequently, at high Péclet numbers both the spatial deviation of the velocity as well as the diffusion determine the dispersive transport.

Table 1. Selected descriptors of the porous structure of the in-line cylinders. The diffusive tortuosity values presented were computed with the MVA approach.

ε [−]	τ_{xx}^{-1} [−]	τ_{xx} [−]	τ_{h_x} [−]
0.37	0.545	1.83	1.0196
0.50	0.649	1.54	1.0192
0.80	0.833	1.20	1.0189
0.95	0.952	1.05	1.0143

MVA vs. BD. As shown in Figs. 2 and 3, the results delivered by both methods are in good agreement with each other. Only in the case of the radial dispersion at $\varepsilon = 0.37$ (see Fig. 3) a slight discrepancy was observed. This might be explained by the fact that the geometry used in the BD approach shows a higher diffusive tortuosity of 1.88, as opposed to the tortuosity value of 1.54 obtained in the MVA approach. Another explanation may be the difference in surface smoothness due to the different discretization methods used in each approach.

Both approaches are able to deal with homogeneous porous media. In terms of computational performance MVA is significantly faster, since only one set of differential equations is solved for each calculation of the dispersion coefficient at a given Péclet number. Just as expected from a deterministic differential equation, MVA always delivers the same numeric results which are subject to classical numerical errors, e.g. involving the number of nodes of the mesh. In

contrast, when using BD we have to solve the equation of motion for each particle of the ensemble considered in the calculations. We compute the trajectories
of 10^6 tracers so that the fluctuations in the solution do not alter the first three
significant digits of the dispersion coefficient and diffusive tortuosity values. Furthermore, the BD approach can handle mass transport in heterogeneous porous
media, since no representative periodic unit cell is needed. This method is ideal
for small systems where boundary effects cannot be neglected. In addition, when
the BD method is used in combination with the LBM, we get a mesh-free simulation approach, which can be of great advantage for complex geometries.

3.2 Mass Transport in Heterogeneous Porous Media: Confinement Effects

This section outlines the dispersion in heterogeneous porous structures consisting of confined random sphere packings. Confinement effects on mass transport are particularly important in column chromatography and can be split
into two effects. The first one is directly related to the highly ordered structure
present in the immediate vicinity of the wall, as opposed to the highly disordered configuration present in the bulk of the packing, known as random close

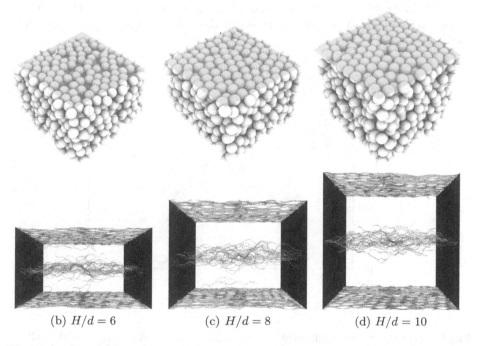

    (b) $H/d = 6$ (c) $H/d = 8$ (d) $H/d = 10$

Fig. 4. At the top the sphere packings of finite H/d-ratios are shown. Below, characteristic pathlines close to the wall ($z \approx 0$) and in the middle of the packing ($z = H/2$)
are plotted for each packing, respectively.

packing (RCP) [40]. This effect alters the tortuosity of the pathlines affecting the mechanical dispersion. The second wall effect emerges from the steep velocity gradient induced by the presence of the wall, which in turn has an enhancing effect on the hydrodynamic dispersion. This contribution to dispersion is also known as trans-column dispersion in literature [41].

The investigated packings were generated numerically with different ratios of packing height to sphere diameter H/d (see Fig. 4). The dispersion coefficients at different Péclet numbers were computed using the BD approach. The packings were generated using the open-source software RCP provided by Desmond et al. [42]. In contrast to the previous case, here the effective porosities were set approximately constant at 0.375 ± 0.005. Each diameter was spatially resolved with 35 lattice cells. The results of the dispersion coefficients are shown in Figs. 5 and 6.

Confinement Effect. The confinement effect on the arrangement of the spheres can be seen in Fig. 4. Here, as mentioned above, the walls constrain the spheres to a highly ordered arrangement. The influence of the walls on the structure of the medium is in the range of 4 to 5 sphere diameters [43]. Outside this region, the characteristic structural disorder of a RCP configuration prevails. This effect can be seen in the porosity profile along the axis perpendicular to the walls (here referred to as the z-axis) of the generated packings: the porosity profiles are supported by the projections of the sphere centers right below each profile in Fig. 8.

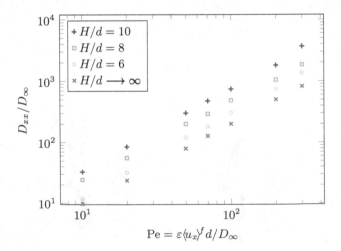

Fig. 5. Axial dispersion coefficients as a function of the Péclet number in the sphere packings considered.

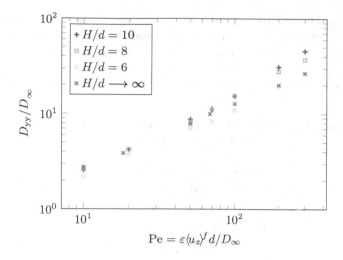

Fig. 6. Radial dispersion coefficients as a function of the Péclet number in the sphere packings considered.

The narrowest packing examined in this study has a height of 6 sphere diameters which is why the entire packing is characterized by an ordered structure. The case of an unconfined sphere packing, i.e. $H/d \longrightarrow \infty$, is modelled by introducing periodic boundary conditions. The porosity profile of the narrowest packing shows pronounced fluctuations that are regularly spaced along its entire extent in the z-axis (see Fig. 8a), whereas the porosity fluctuations of the periodic packing are smaller and homogeneously distributed, see Fig. 8d.

To illustrate how the difference in structural disorder between these two extreme cases affects the convective motion of the tracers, we plotted pathlines near the wall at $z \approx 0$ and those in the middle of the packing at $z = H/2$ (see Figs. 4 and 7). The pathlines of the periodic packing ($H/d \longrightarrow \infty$) were also graphically depicted in Fig. 7e.

The packings with H/d ratios of 8 and 10 have different volume proportions of ordered and disordered pore regions. The presence of two pore regions with different structures in the same packing leads to the development of two significantly different flow conditions in each region, which results in strongly heterogeneous velocity profiles; heterogeneous not only in terms of magnitude but in the tortuosity of the stream paths along the pore network. This in turn leads to a strong dispersion of the diffusing tracers, i.e., to large dispersion coefficients. Both, the infinitely extended packing and the narrowest packing have mainly only one characteristic pore region, i.e., a lesser degree of heterogeneity, which explains their lower dispersion coefficients (see Figs. 5 and 6). Similar results were obtained in the experimental work of Bruns et al. 2013, for column chromatography with varying ratio of the column diameter to the mean particle diameter [41].

Diffusive vs. Hydraulic Tortuosity. Regarding the selected descriptors, the diffusive tortuosity remains approximately constant around 1.4 for all the packings considered, see Table 2. This is a direct consequence of the fact that the effective porosity was set as constant. These results show the strong correlation of the diffusive tortuosity with the effective porosity of the sphere packings.

Table 2. Selected descriptors of the porous structure of the sphere packings. The tortuosity values have a standard deviation of 0.03.

H/d [−]	τ_{xx}^{-1} [−]	τ_{xx} [−]	τ_{h_x} [−]
6	0.694	1.44	1.2328
8	0.694	1.44	1.2455
10	0.694	1.44	1.2486
∞	0.685	1.46	1.2723

In contrast, the hydraulic tortuosity increases slightly with increasing packing height. However, as the descriptors are average (effective) values, neither of them is suitable to capture the contribution of the heterogeneity in the structure to the dispersive mass transport. As opposed to the homogeneous case, where the hydraulic tortuosity directly correlates with the dispersion coefficient, here an indirect correlation between these two parameters is observed, meaning that the porous structure with the largest value of hydraulic tortuosity showed the smallest dispersion coefficient, namely the unconfined sphere packing with $H/d \longrightarrow \infty$.

From the results it can be deduced that in cases where heterogeneity is strongly pronounced in the structure of a porous medium, effective descriptors are not well suited to describe the dispersive mass transport in the medium. Instead, a spatial distribution or a probability distribution of the local hydraulic tortuosity could be used, whose standard deviation captures the degree of heterogeneity of a porous medium. A spatial distribution can be achieved by applying Eq. 17 to each plane along a selected axis, from which a tortuosity profile could be represented (analogous to the porosity profiles in Fig. 8). A probability distribution could be attained by calculating the tortuosity and occurrence of individual pathlines. These pathlines can be obtained e.g. using a particle tracking method.

According to the results, the degree of heterogeneity of a porous medium governs dispersion against the (effective) diffusive and hydraulic tortuosity in confined sphere packings at high Péclet numbers.

In a qualitative analysis, we can take the values of the narrowest confined packing $H/d = 6$ and the unconfined packing $H/d \longrightarrow \infty$ as characteristic tortuosity values of the wall and bulk regions of a packing, respectively. Here, the wall region is to be understood as the region where the ordered configuration induced by the wall is present, whereas the bulk region is the region in which the RCP configuration is found. So we can express the hydraulic tortuosity of

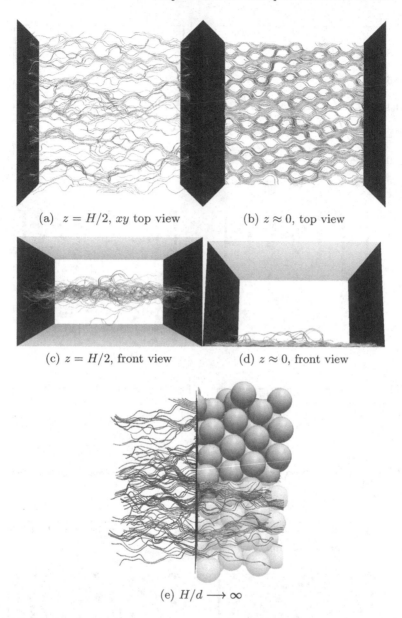

(a) $z = H/2$, xy top view   (b) $z \approx 0$, top view

(c) $z = H/2$, front view   (d) $z \approx 0$, front view

(e) $H/d \longrightarrow \infty$

Fig. 7. (a)–(d) show selected views of the pathlines in the sphere packing with $H/d = 6$: close to the wall at $z \approx 0$ in (b) and (d), and in the middle of the packing at $z = H/2$ in (a) and (c). Additionally in (e), a portion of the periodic sphere packing and its related pathlines are presented.

the larger confined packings as the volume-weighted average of the characteristic tortuosity values. In this way, the degree of heterogeneity is related to the volume fractions of each characteristic region, where a volume fraction of one or zero indicates a homogeneous medium.

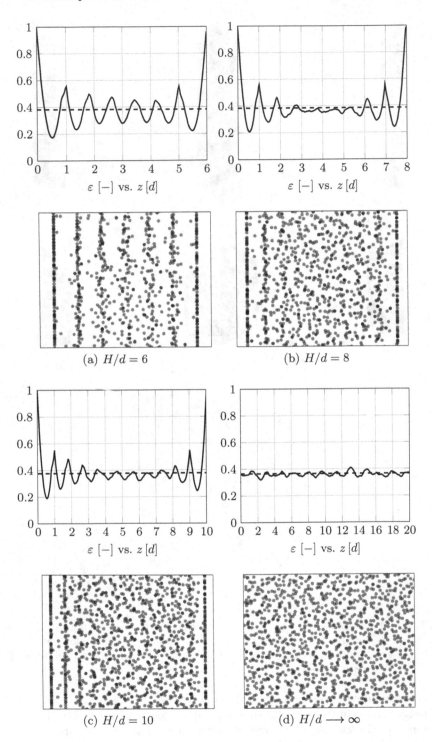

Fig. 8. Porosity profiles along the z-axis and projections of the sphere centers on a plane perpendicular to the walls, also along the z-axis from left to right.

4   Conclusions

With the presented approaches we were able to determine and discuss the transport properties of the homogeneous and heterogeneous porous media considered. In the structured geometry of the homogeneous case, a clear correlation of the dispersive transport properties of the media with their tortuosity and effective porosity values can be seen, as opposed to the heterogeneous case, where there is no direct correlation between the considered parameters. The reason is attributed to the fact that the used descriptors are effective values not capable of capturing the dispersive effect induced by the degree of heterogeneity in the microscopic porous structure. Instead of effective parameters, the use of a local hydraulic tortuosity is proposed, the implementation of which will be left to future work.

In addition, the MVA and BD approaches were compared in the section of homogeneous porous media. Here, the method of volume average outperformed Brownian dynamics in terms of both precision and computational performance.

Acknowledgment. The authors appreciate the support for this project by the North-German Supercomputing Alliance (HLRN) as well as by the Lower Saxony Ministry of Science and Culture within the PhD program "Self-organizing multifunctional structures for adaptive high performance light-weight constructions". The framework of this coordinated program is the "Campus for Functional Materials and Functional Structures", an institution of the Clausthal University of Technology (TUC) in collaboration with the DLR, German Aerospace Center, in Braunschweig, the BAM, Federal Institute for Material Testing, in Berlin, and the Technical University of Braunschweig (TU BS).

References

1. Ghanbarian, B., Hunt, A.G., Ewing, R.P., Sahimi, M.: Tortuosity in porous media: a critical review. Soil Sci. Soc. Am. J. **77**(5), 1461 (2013)
2. Carman, P.C.: Fluid flow through granular beds. Chem. Eng. Res. Des. **75**, S32–S48 (1937)
3. Bear, J.: Dynamics of Fluids in Porous Media, vol. 1. American Elsevier Publishing Company, New York (1972)
4. Berg, C.F.: Permeability description by characteristic length, tortuosity, constriction and porosity. Transp. Porous Media **103**(3), 381–400 (2014). https://doi.org/10.1007/s11242-014-0307-6
5. Niya, S.M.R., Selvadurai, A.P.S.: A statistical correlation between permeability, porosity, tortuosity and conductance. Transp. Porous Media **121**(3), 741–752 (2017). https://doi.org/10.1007/s11242-017-0983-0
6. Whitaker, S.: The Method of Volume Averaging. Theory and Applications of Transport in Porous Media, vol. 13. Kluwer Academic Publishers, Dordrecht (1999)
7. Valdes-Parada, F.J., Aguilar-Madera, C.G.: Upscaling mass transport with homogeneous and heterogeneous reaction in porous media. Chem. Eng. Trans. **24**, 1453–1458 (2011)
8. Lugo-Méndez, H.D., Valdés-Parada, F.J., Porter, M.L., Wood, B.D., Ochoa-Tapia, J.A.: Upscaling diffusion and nonlinear reactive mass transport in homogeneous porous media. Transp. Porous Media **107**(3), 683–716 (2015). https://doi.org/10.1007/s11242-015-0462-4

9. Heidig, T., Zeiser, T., Schwieger, W., Freund, H.: Ortsaufgelöste Simulation des externen Stofftransports in komplexen Katalysatorträgergeometrien. Chem.-Ing.-Tech. **86**(4), 554–560 (2014)
10. Khirevich, S., Daneyko, A., Höltzel, A., Seidel-Morgenstern, A., Tallarek, U.: Statistical analysis of packed beds, the origin of short-range disorder, and its impact on eddy dispersion. J. Chromatogr. A **1217**(28), 4713–4722 (2010)
11. Koku, H., Maier, R.S., Schure, M.R., Lenhoff, A.M.: Modeling of dispersion in a polymeric chromatographic monolith. J. Chromatogr. A **1237**, 55–63 (2012)
12. Rusinque, H., Brenner, G.: Mass transport in porous media at the micro- and nanoscale: a novel method to model hindered diffusion. Microporous Mesoporous Mater. **280**, 157–165 (2019)
13. Hornung, U. (ed.): Homogenization and Porous Media. Interdisciplinary Applied Mathematics, vol. 6. Springer, New York (1997). https://doi.org/10.1007/978-1-4612-1920-0
14. Gray, W.G., Miller, C.T.: Thermodynamically constrained averaging theory approach for modeling flow and transport phenomena in porous medium systems: 1. Motivation and overview. Adv. Water Resour. **28**(2), 161–180 (2005)
15. Kirby, B.J.: Micro- and Nanoscale Fluid Mechanics: Transport in Microfluidic Devices. Cambridge University Press, Cambridge (2010)
16. Leal, L.G.: Advanced Transport Phenomena: Fluid Mechanics and Convective Transport Processes. Cambridge Series in Chemical Engineering. Cambridge University Press, Cambridge (2010)
17. Perumal, D.A., Dass, A.K.: A review on the development of lattice Boltzmann computation of macro fluid flows and heat transfer. Alex. Eng. J. **54**(4), 955–971 (2015)
18. Shan, X.: Lattice Boltzmann in micro- and nano-flow simulations. IMA J. Appl. Math. **76**(5), 650–660 (2011)
19. Alnæs, M.S., Logg, A., Ølgaard, K.B., Rognes, M.E., Wells, G.N.: Unified form language: a domain-specific language for weak formulations of partial differential equations. ACM Trans. Math. Softw. **40**(2) (2014)
20. Alnæs, M.S., et al.: The FEniCs project version 1.5. Arch. Numer. Softw. **3**(100) (2015)
21. He, X., Luo, L.S.: Theory of the lattice Boltzmann method: from the Boltzmann equation to the lattice Boltzmann equation. Phys. Rev. E - Stat. Phys. Plasmas Fluids Relat. Interdiscip. Top. **55**(6), 6811–6820 (1997)
22. Succi, S.: The Lattice Boltzmann Equation for Fluid Dynamics and Beyond (Numerical Mathematics and Scientific Computation). Oxford University Press, Oxford (2001)
23. Wolf-Gladrow, D.A.: Lattice Gas Cellular Automata and Lattice Boltzmann Models. LNM, vol. 1725. Springer, Heidelberg (2000). https://doi.org/10.1007/b72010
24. Aris, R.: On the dispersion of a solute in a fluid flowing through a tube. Proc. R. Soc. A Math. Phys. Eng. Sci. **235**(1200), 67–77 (1956)
25. Koponen, A., Kataja, M., Timonen, J.: Tortuous flow in porous media. Phys. Rev. E **54**, 406–410 (1996)
26. Duda, A., Koza, Z., Matyka, M.: Hydraulic tortuosity in arbitrary porous media flow. Phys. Rev. E **84**, 036319 (2011)
27. Schuster, H.G., Just, W.: Deterministic Chaos. Wiley-VCH, Weinheim (1985)
28. Volpe, G., Volpe, G.: Simulation of a Brownian particle in an optical trap. Am. J. Phys. **81**(3), 224–230 (2013)
29. Bian, X., Kim, C., Karniadakis, G.E.: 111 years of Brownian motion. Soft Matter **12**(30), 6331–6346 (2016)

30. Rusinque, H., Fedianina, E., Weber, A., Brenner, G.: Numerical study of the controlled electrodeposition of charged nanoparticles in an electric field. J. Aerosol Sci. **129**, 28–39 (2018)
31. Boltzmann, L.: Vorlesungen über Gastheorie (translated into English as "Lectures on Gas Theory"). Vorlesungen über Gastheorie. J. A. Barth (1898)
32. Deen, W.M.: Hindered transport of large molecules in liquid-filled pores. AIChE J. **33**(9), 1409–1425 (1987)
33. Zhang, G., Schlick, T.: Implicit discretization schemes for Langevin dynamics. Mol. Phys. **84**(6), 1077–1098 (1995)
34. Hoze, N., Holcman, D.: Statistical methods for large ensemble of super-resolution stochastic single particle trajectories. bioRxiv, p. 227090, November 2017
35. Jones, D.K.: Diffusion MRI. Oxford University Press, Oxford (2010)
36. Dechadilok, P., Deen, W.M.: Hindrance factors for diffusion and convection in pores. Ind. Eng. Chem. Res. **45**(21), 6953–6959 (2006)
37. Evans, R., Dal Poggetto, G., Nilsson, M., Morris, G.A.: Improving the interpretation of small molecule diffusion coefficients. Anal. Chem. **90**(6), 3987–3994 (2018)
38. Xiao, J., Chen, X.D.: Multiscale modeling for surface composition of spray-dried two-component powders. AIChE J. **60**(7), 2416–2427 (2014)
39. Dvořák, P., Šoltésová, M., Lang, J.: Microfriction correction factor to the Stokes-Einstein equation for small molecules determined by NMR diffusion measurements and hydrodynamic modelling. Mol. Phys. **117**, 868–876 (2018)
40. Shalliker, R.A., Broyles, B.S., Guiochon, G.: Physical evidence of two wall effects in liquid chromatography. J. Chromatogr. A **888**(1–2), 1–12 (2000)
41. Bruns, S., Franklin, E.G., Grinias, J.P., Godinho, J.M., Jorgenson, J.W., Tallarek, U.: Slurry concentration effects on the bed morphology and separation efficiency of capillaries packed with sub-2 μm particles. J. Chromatogr. A **1318**, 189–197 (2013)
42. Desmond, K.W., Weeks, E.R.: Random close packing of disks and spheres in confined geometries. Phys. Rev. E **80**, 051305 (2009)
43. Maier, R.S., Kroll, D.M., Bernard, R.S., Howington, S.E., Peters, J.F., Davis, H.T.: Hydrodynamic dispersion in confined packed beds. Phys. Fluids **15**(12), 3795–3815 (2003)

Generative Design Solutions for Free-Form Structures Based on Biomimicry

Gaurab Sundar Dutta(✉), Leif Steuernagel, and Dieter Meiners

Institute of Polymer Materials and Plastic Engineering, Clausthal University of Technology, 38678 Clausthal-Zellerfeld, Germany
{gaurab.sundar.dutta,leif.steuernagel,
dieter.meiners}@tu-clausthal.de

Abstract. Conventional engineering design approach is to solve a structure for its strength. On the other hand, natural shapes and structures are more free-form, using stiffness and flexibility whenever and wherever necessary. Form-finding approaches for structural design draws inspiration from different trades of nature. In this work, an attempt is made to generate free-form structures using biological forms like growth of a plant. Generative design strategy within constrained design space was developed, and behavior of such solutions under different loading are discussed. Thus a structure is evolved (and adapted) *via* bottom-up approach, and not constructed (and optimized) as top-down process.

Selected examples show the robustness and flexibility of the structures depending on the application. The testing examples also enable the designer to improve the problem statement along with the solution sets. A smooth back and forth parametric modeling process facilitates a smooth interaction between design and analysis, which is essential in form-finding approaches. A free-form self-organized shape would enable the algorithm to directly interact with printing techniques, easing manufacturing.

Keywords: Generative design · Free-form · Biological form · Evolutionary algorithm · Self-organized structure · Parametric model

1 Introduction

Design, depending on the area of application, differs in its definition. In structural engineering, it is the process of investigation of a given structure for stability and strength; under the prescribed loading scenario. The objective there is to produce a structure, capable enough to transmit and resist the applied loads without failure, within the intended life, while *design* as the trade of an architect; is mostly an *ill structured problem*. The lack of complete problem definition, and constraints makes the problem ill poised. While being *creative* is one prime objective, the stability of the structure under intended constraints are also equally important trades [1]. Over the past few decades, there has been a significant emphasis on interdisciplinary and sustainable engineering projects. This in turn led to collaborative projects such as facade design and evolved building forms, which demand attention of both engineering and architectural field of knowledge [2–4].

© Springer Nature Switzerland AG 2020
N. Gunkelmann and M. Baum (Eds.): SimScience 2019, CCIS 1199, pp. 122–134, 2020.
https://doi.org/10.1007/978-3-030-45718-1_8

Consequently, design in contemporary state-of-the-art has slowly evolved into a complex interconnected network of search processes where various solutions of varying efficiencies are generated, along with better understanding of the problem domain itself [5]. The evolutionary search process can be seen as advancement and backtracking of solution strategies in order to adapt a solution rather than optimizing it, thus resulting in a range of solutions, generated under different combination of independent constraints, thus the name: *generative design*.

Generative design derives its inspiration from Nature. Nature following the laws of morphogenesis [6] adapts shapes and forms attaining self-organization according to the requirements, over the period of time (see Fig. 1). For years, researchers have been trying to develop *design* as search paradigm in laboratories and computational domain [7], and thus evolving the idea of generative design. In current times, with the advancements of computation and mathematical techniques, it is now possible to visualize different forms virtually, realizing the same with existing 3D printing technologies.

Fig. 1. (left) 'Tree and roots' [8], (right) 'Culmann's crane' [9].

Computational morphogenesis in engineering is to find best possible shapes and material distributions of a structure, in response to external constraints, mostly to minimize weight, deflection and mechanical stress. In many ways, the objectives are similar in animal and plant evolution as well [10]. For plants, growth in optimal path ensures maximum accumulation of resources like light or minerals with minimum work done; while for animals, efficient growth in limbs and bones are to minimize energy consumption in response to external forces encountered during walking or running. Present work explores the very idea of self-organized structures and generative design evolving simple free form shapes as examples. These shapes further act as building blocks creating efficient complex geometries.

A prime objective here is to find the effect of orientation in the geometry of a structure using bottom-up approach [11], under given boundary constraints. The results can encourage a tailored fibre placement process. In recent years, there has been an increasing interest and advancements in continuous fibre reinforced lightweight structures, owing to their large stiffness to weight ratios, improved fatigue life, steadiness under compressive force, and good designability [12]. Fibres are transversely isotropic in nature. Hence, exploring *optimal* path would eventually result in an efficient structure, exploiting both material and geometrical properties.

2 Sample Problem and Design Strategies

Initial problem set-up is a two-point problem, where one is assigned as fixed *base point*, and the other acts as the *load point* with an external load input. A very obvious solution in this case would be a straight line, joining these two points.

Here, the length of the member is relaxed, such that the curve thus generated adjusts itself according to the given conditions, and thus would refer to the *optimal* path. The problem is thus translated into finding the best suited curve and consequently comparing its performance in contrast with the straight-line solution. For simplicity, only one force in the negative direction of Z-axis is considered for now.

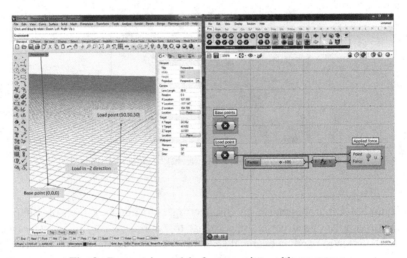

Fig. 2. Parametric model of a two-point problem set-up.

A parametric set-up of the problem (see Fig. 2) was designed using Grasshopper 3D [13] visual programming tool within Rhino [14] environment. Parametric set-up provides flexibility to the designer in incorporating and changing component information; a key essence of design as a search process. In order to structure the design exploration strategy, a space is required within which the search is kept confined. This is referred to as *design space*. Design space can be of any shape depending on the problem constraint, creative preferences or any other external element. In this case, a cubic volume with base and load points at two extreme diagonals is considered (see Fig. 3). Hence the domain for each new point comprising the solution will be confined within two extremities.

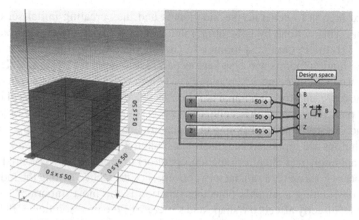

Fig. 3. Parametric model of design space with XYZ-inputs.

2.1 Evolutionary Programming and Plant Growth Mechanics

Evolutionary Programming (EP) as an optimization tool works much similar to Genetic Algorithms (GA) and represents a whole range of probabilistic algorithms which are parallel in nature and are inspired by natural selection process [15, 16]. While deterministic algorithms such as conjugate gradient or gradient descent works in a specific direction, these algorithms simultaneously search for optima in all available directions. Thus they are less prone to be trapped in local minima and depending on the test cases, perform better than some standard deterministic tools to find convergences. They work with simple inputs as foundation units and tend to build the solution structure upon them. These inputs vary depending upon the problem set-up and referred to as phenomes/phenotypes in EPs. On the contrary, GAs refer to similar inputs as genomes/genotypes, which are basically binary representation of their EP counterparts. Consequently, EP solutions are visually better reportative than GA solutions, with a tradeoff in computational time. Based on the input information, a gene pool of potential solution is created for testing. This set of solutions is called one *generation* and each potential solution is called one *individual*. Visualizing each individual during generation was a prime reason behind choosing EP over GA in this work.

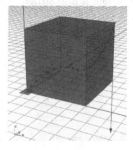

Fig. 4. An intermediate solution or *individual*.

The next step is to generate the individuals and thus creating initial set of possible solutions. This is done by generating random points inside the design space using GhPython [17], the Python [18] toolbox in Grasshopper environment. The number of intermediate random points was also kept as parametric input for flexibility. The points along with two extreme points when joined as a polyline, is referred here as phenotype or intermediate representative individual (see Fig. 4). Once prepared, these sets of individuals are then passed through a filter, referred to as cost-function/fitness function/objective function and given a rank according to their performance. Based on the ranks, best performing solutions are separated and much like a natural evolution process are mixed up (crossover operator) and altered (mutation operator) to create a new potential solution set. The process goes on till a satisfactory convergence is reached among the individuals.

Thus, constructing a suitable objective function is the most important part of the algorithm as it directly affects the shape and structure of the final solution. CPU time for calculation also depends on the complexity of the objective function and amount of noise in the data set. The objective function also suggests what kind of solution the designer is seeking and thus expanding the problem formulation, its application and possible *best-fit* solutions.

Fig. 5. Growth of plant towards light source.

In plants, the mechanism of formation is understood to be driven by the concentration of auxins, a hormone that regulates new cellular growth and coordinates the emergence of the plant's geometry precisely towards available resources [19]. Two key resources for tree are light and minerals. Consequently the growth of auxins toward each of them is in the form of branch and root. In this work growth of tree in branching direction will be discussed. This is primarily due to light response or commonly known as *Phototropism*. Essentially, the auxins which are exposed to light at any instance get saturated, thus enabling the shape to bend towards light direction (see Fig. 5). The idea here is to mimic this phenomenon find a way to translate it into structural form.

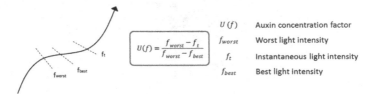

$$U(f) = \frac{f_{worst} - f_t}{f_{worst} - f_{best}}$$

$U(f)$	Auxin concentration factor
f_{worst}	Worst light intensity
f_t	Instantaneous light intensity
f_{best}	Best light intensity

Fig. 6. Plant growth algorithm.

In the field of botany and biochemistry, light response curves have been experimented using mathematical models which very recently been adapted in neural network domain [20]. The formulation of the model relates the auxin concentration factor with relative light intensity at any instant as a hyperbolic function (see Fig. 6).

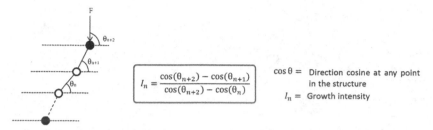

$$I_n = \frac{\cos(\theta_{n+2}) - \cos(\theta_{n+1})}{\cos(\theta_{n+2}) - \cos(\theta_n)}$$

$\cos \theta$ = Direction cosine at any point in the structure

I_n = Growth intensity

Fig. 7. Structural analogy to plant growth algorithm.

Here, a parallel is drawn between tree auxins – light relationship, with orientation of structure in presence of external load. The light intensity factor is replaced with the direction cosines associate with each point vectors of the proposed solution, worst value referring to the *base point* and best referring to the *load point*, which eventually is an external input associated with the applied load.

The resulting value is referred to as *growth intensity* (see Fig. 7) of the structure at any particular point. Accumulation of this value for all three orthogonal directions would form the objective function, where the goal would be to minimize this accumulated value. An additional constraint for form generation is realized by external relaxed length input. This input is evaluated from the results obtained in the method described next.

2.2 Particle Spring Method

Before discussing results of the previous section, the Particle-spring method, an alternative existing approach is discussed. This will in turn serve as a validation for the plant growth inspired solution sets.

The Particle-spring method is a well-established tool in animation industry to create physical simulations for the likes of hair and waves. This draws inspiration from form-finding techniques described in the previous section as well and lately been used in designing free form architectural structures.

The mechanism, as the name suggests is constituted of lumped masses called particles, connected by linear elastic spring. The stiffness of the springs allows the shape to change under applied load (see Fig. 8) and eventually stabilizes at the rest length [21].

For this exercise, Kangaroo Physics toolbox (Grasshopper) [22] is used to assign relevant properties to the individuals and extract kinetic energy as output. Like other parametric tools, Kangaroo Physics also takes parameters like segmentation, stiffness, damping, particle-mass and number of iteration (for solver) as variable inputs. Segmentation is kept moderately low to save computational time. Stiffness values are calculated

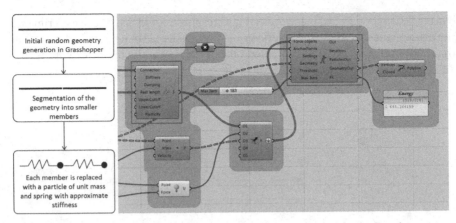

Fig. 8. Particle spring method flowchart and programming modules in Grasshopper.

from material properties, considering each segment as small cantilevers, and are kept sufficiently high compared to rest-length. Damping and mass were kept as unity considering completely elastic system.

The number of iterations was set to be a relatively high number, signifying deformation after a sufficient period of loading time. At this point, energy of the system, an output from the solver is defined as the fitness/objective function, while the random points comprising the curve are taken as phenotypes for design evaluation. Galapagos [23], an inbuilt evolutionary computing solver inside Grasshopper environment is used for this set-up (see Fig. 9).

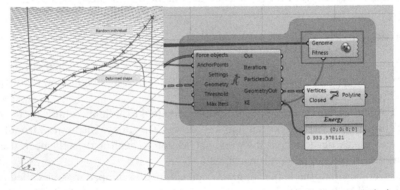

Fig. 9. Deformed shape of a randomly generated solution with Galapagos optimizer.

2.3 Results Comparison and Discussion

For both the design strategies described above, the external load value is considered to be 100N acting in negative Z-direction. Material properties of ABS [24] are taken into consideration for all the calculations including stiffness values and further finite element analysis. The Galapagos solution yields a curve aligning itself in the direction of load.

The length of this curve serves as the additional input in the plant growth algorithm. The resulting curves (see Fig. 10) are found to be in good agreement with each other.

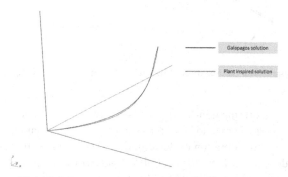

Fig. 10. Solution curves from two design strategies.

Thus, two independent systems of solution were explored complementing each other under similar constraints and conditions. Galapagos produce a global minimum, primarily depending on the number of segmentations, with other inputs. An increment in the segmentation number results in a curvature of larger length, with an increase in computational time. This length when provided as an input to plant growth operator, converges to the Galapagos solution. The orientation of structure in the direction of load is true for any force definitions. The advantage of this method is that the length, an integral part of the structure and material consumption is set-up as an input variable. This empowers designer to create shapes for specific material requirements and need, thus resulting in a generative design system.

3   Usage and Applications

At this conjuncture, the essential query was to find the properties and use of this kind of oriented structures. Thus, a series of examples and test cases were explored within the simulation domain, generated via different mathematical operations on the curved shapes and their linear counterparts. This exercise led to a better understanding of not only the solutions but the essential problem domain itself.

3.1   Design Problem Enrichment

The first test was conducted in ABAQUS [25] finite element solver by exporting a representative volume geometry from Rhino for line and generated-curve structure, and subjecting them under exactly similar boundary conditions described earlier. A non-linear simulation yields different phases of structural deformation each of the geometries undergoes (see Figs. 11 and 12).

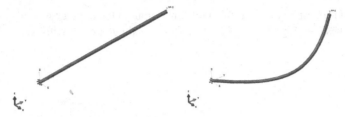

Fig. 11. Finite element analysis set-up in ABAQUS for (left) linear and (right) curved models.

The reason for performing non-linear analyses was to observe deformed shapes in a more realistic condition where load is normally applied via external weights and follows the structure for the whole duration of deformation. The line structures being compression driven, provide more stiffness to the applied load before bending is encountered and eventually elongated along the direction of load. On the other hand, the curved geometry undergoes large deformation initially at the bending mode, before compression and elongation subsequently take over.

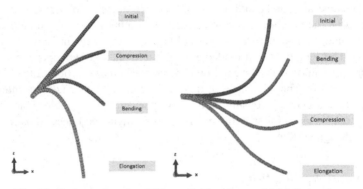

Fig. 12. Deformed modes of linear (left) and curved (right) structures.

The key takeaway from this test was that, a structure aligned in the direction of the load offers less resistance to the load, consequently becoming flexible in that specific direction.

The next step was to observe the effect of in-plane deformation on the shape of the structure. To examine this, both the extreme points of the structure are kept fixed and a load in the XY-plane is applied at an intermediate point. The location of the loading point is kept consistent for both the structures. The deformed shape comparison (see Fig. 13) suggests that the linear member suffers large deformation and higher stress values in contrast to the curved counterpart. It can easily be observed that in a line model the structure suffers significant bending at the loading point, while the curved structure is elongated first to negate the effect of loading.

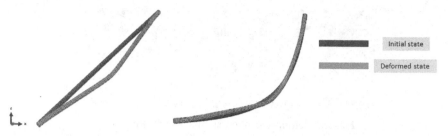

Fig. 13. Initial and final state of deformation for line (left) and curve (right) under XY-load.

This was indeed an encouraging result, explaining more about the usefulness of the shape and expanding domain of the problem. Major take away here was that if a structure is designed by aligning its growth in the direction of a given load, its shape exhibits strength in other orthogonal directions, with extremities being fixed.

3.2 Explanatory Examples

Based on the findings from initial tests, a number of applications were thought of. A couple of them are mentioned here, explaining the use of strength and flexibility of the shape.

First example is a surface of revolution, using both line and curve model. Inspiration of this idea is drawn from tensile structures which are often used as roof-tops and tent-like structures. Karamba 3D [26], a parametric finite element solver inside Grasshopper domain is used to analyze both the models for shell deformation. The loading position, value and direction are again kept constant for both the models, while top and bottom of the surfaces are kept fixed (see Fig. 14).

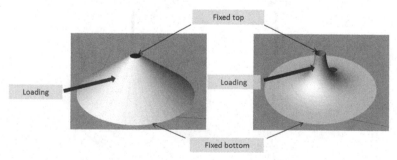

Fig. 14. Simulation set-up of shell structures generated from (left) line and (right) curve geometry for planar load.

The deformation (see Fig. 15) replicates results from solid geometry with minimal effect on the curved geometry. This can even be observed for geometries generated by patched surface of revolution as well (see Fig. 16).

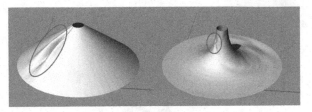

Fig. 15. Deformed configurations of shell structures.

Fig. 16. Deformation in shell structure patches with planer load and fixed boundaries.

Second example is an attempt to exploit the flexibility feature of the load oriented curved shape compared to the line structure. Press-fit samples are modeled by mirroring the line and curve shapes about a central axis at 90, 180 and 270° and are assigning suitable geometrical properties (see Fig. 17) for ABAQUS simulation. Press-fit design was preferred for the ease in manufacturability via 3D printing process. A compressive displacement load is applied to each of the samples, while the bottom of the frame being fixed.

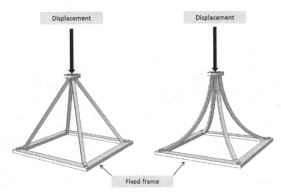

Fig. 17. Framed compression structures generated from (left) line and (right) curve samples.

From deformed shape configuration comparison (see Fig. 18) it is evident that the line structure undergoes multiple modes of bending, while the curved structure compresses smoothly for the same amount of displacement constraint. Stress and reaction force values of the line-structure are also consequently higher than the curved counterpart.

Fig. 18. Deformation in (left) linear and (right) curved samples.

It is interesting to note that, with the curved structure, a spring-like mechanism plays a significant role providing the necessary relaxation, while a linear structure being stiff enough is not able to exhibit flexibility.

4   Conclusion

Design as a search process is a promising method, which can lead to a range of effective solutions under certain constraints. The process of evolution when translated into a parametric model, allows the designer to realize, alter and correlate different independent variables involved in a complex design process. Thus a *generative* range of solutions are evolved, which in turn enable better understanding and extension of the problem domain itself. In this work, a bio-inspired model was formulated to evolve free-form structures by mimicking plant growth mechanisms. An evolutionary algorithm, inspired by natural selection process was adopted to search for the solution individuals. Importance of design space in evolutionary computation and its effect was emphasized. The solutions were compared with standard evolutionary solvers in the parametric modeling domain. Further, generated structures were analyzed under different environments to understand their geometric properties. The knowledge gained, were used to extend the problem set into more realistic example set-ups. Each example was compared with a similar structure generated *via* line-model. The results were found to be encouraging. With these results it can be safely concluded that a free-form shape mimicking a biological process results in an efficient curve with multiple advantages. A number of use cases thus can be considered as future work. Moreover, depending on the problem constraints, the shape of the path can be directly integrated with available manufacturing processes, enabling digital manufacturing specifically with transversely isotropic materials like carbon fibres.

Acknowledgements. The work has been funded by the Lower Saxony Ministry of Science and Culture as a part of the Ph.D. program *Campus FWS*.

References

1. Simon, H.: The structure of ill structured problems. Artif. Intell. **4**, 181–201 (1973)
2. Nagel, R.L., Pappas, E.C., Pierrakos, O.: On a vision to educating students in sustainability and design–the James Madison University School of Engineering approach. Sustainability **4**(1), 72–91 (2012)

3. El-Alfy, A.: Design of sustainable buildings through value engineering. J. Build. Apprais. **6**(1), 69–79 (2010)
4. Kovacic, I., Filzmoser, M., Denk, F.: Interdisciplinary design: influence of team structure on project success. Procedia – Soc. Behav. Sci. **119**, 549–556 (2014)
5. Pahl, G., Frankenberger, E., Badke-Schaub, P.: Historical background and aims of inter-disciplinary research between Bamberg Darmstadt and Munich. Des. Stud. **20**(5), 401–406 (1999)
6. Hensel, M., Menges, A., Weinstock, M.: Introduction to emergence: morphogenetic design strategies. In: Carpo, M. (ed.) The Digital Turn in Architecture 1992–2012, 1st edn. Wiley, Chichester (2013)
7. Woodbury, F.R.: Searching for designs: paradigm and practice. Build. Environ. **26**(1), 61–73 (1991)
8. Bejan, A.: Shape and Structure, from Engineering to Nature. Cambridge University Press, Cambridge (2000)
9. Thompson, D.: On Growth and Form, 12th edn. Cambridge University Press, Cambridge (1992)
10. Aage, N., Andreassen, E., Lazarov, B.S., Sigmund, O.: Giga-voxel computational morpho-genesis for structural design. Nature **550**, 84–86 (2017)
11. Aziz, M.S., El Sheriff, A.Y.: Biomimicry as an approach for bio-inspired structure with the aid of computation. Alex. Eng. J. **55**(1), 707–714 (2016)
12. Hou, Z., Tian, X., Zhang, J., Li, D.: 3D printed continuous fibre reinforced composite corrugated structure. Compos. Struct. **184**, 1005–1010 (2018)
13. Grasshopper 3D. https://www.grasshopper3d.com/. Accessed 01 Aug 2019
14. Rhinoceros 3D. https://www.rhino3d.com/. Accessed 01 Aug 2019
15. Goldberg, D.E.: Genetic Algorithms in Search, Optimization and Machine Learning, 10th edn. Pearson Education Inc., New Delhi (2012)
16. Deb, K.: Multi-Objective Optimization Using Evolutionary Algorithms. Wiley India Pvt. Ltd., New Delhi (2018)
17. GhPython. https://www.food4rhino.com/app/ghpython. Accessed 01 Aug 2019
18. Python programming language. https://www.python.org/. Accessed 01 Aug 2019
19. Tamke, M., Stasiuk, D., Thomsen, M.R.: The rise – material behaviour in generative design. In: ACADIA 2013: Adaptive Architecture [Proceedings of the 33rd Annual Conference of the Association for Computer Aided Design in Architecture (ACADIA)], Cambridge, pp. 379–388 (2013)
20. Cui, Z., Cai, X.: Artificial plant optimization algorithm. In: Yang, X.S., Cui, Z., Xiao, R., Gandomi, A.M., Karamangolu, M. (eds.) Swarm Intelligence and Bio-Inspired Computation Theory and Applications, pp. 351–365. Elsevier, Amsterdam (2013)
21. Killian, A., Ochsendorf, J.: Particle-spring systems for structural form finding. J. Int. Assoc. Shell Spat. Struct. IASS **46**, 77–84 (2005)
22. Kangaroo Physics. https://www.food4rhino.com/app/kangaroo-physics. Accessed 03 Aug 2019
23. Galapagos Evolutionary Computing. https://www.grasshopper3d.com/group/galapagos. Accessed 03 Aug 2019
24. ABS technical data sheet. https://www.innofil3d.com/material-data/abs-technical-data/. Accessed 03 Aug 2019
25. Abaqus 2019. https://www.3ds.com/products-services/simulia/products/abaqus/. Accessed 04 Aug 2019
26. Karamba 3D. https://www.karamba3d.com/. Accessed 04 Aug 2019

Accelerating the Visualization of Gaps and Overlaps for Large and Dynamic Sphere Packings with Bounding Volume Hierarchies

Feng Gu[1]([✉]), Zhixing Yang[2], Michael Kolonko[2], and Thorsten Grosch[1]

[1] Institute of Computer Science, Technical University of Clausthal,
Julius-Albert-Street 2, 38678 Clausthal-Zellerfeld, Germany
`feng.gu@tu-clausthal.de`
[2] Institute of Mathematics, Technical University of Clausthal,
Erzstr. 1, 38678 Clausthal-Zellerfeld, Germany

Abstract. The Collective Rearrangement (CR) algorithm is widely used for simulating packings of spheres to gain insight into many properties of granular matter. The quality of a CR simulation can be judged with a visualization technique by directly visualizing the gaps and overlaps of the spheres with pixel precision in each iteration. This visualization technique is based on an Graphic Processing Unit (GPU) linked list, which requires that information for each pixel of each sphere to be stored in advance and then be sorted for each pixel independently. Such requirements impose restrictions on the scale of sphere packings that can be visualized. Instead, one can use ray tracing to resolve such problems. However, there is no available practical ray traversal algorithm for visualizing overlaps and gaps based on acceleration structures for ray tracing. This paper provides traversal algorithms based on Bounding Volume Hierarchies (BVH) to address this problem and can generally make the visualization process much faster than before and reduce the global memory requirement on the GPU to render larger scenes.

Keywords: Scientific visualization · Ray tracing · Ray traversal

1 Introduction

The waterless particle packing plays an important role in determining important properties of many materials from different fields like concrete, pills and tablets for medical purposes or powders for 3D printing. Simulating and inspecting these packings may therefore help to develop materials with a particularly desirable property. In order to obtain a packing as dense as a real dry mixture in the simulation, one approach named Collective Rearrangement (CR) is applied: Spheres

Supported by organization Simulation Science Center Clausthal-Göttingen and Dyckerhoff Stiftung.

N. Gunkelmann and M. Baum (Eds.): SimScience 2019, CCIS 1199, pp. 135–149, 2020.
https://doi.org/10.1007/978-3-030-45718-1_9

are first placed at random positions in a container that is so small that spheres must overlap. Then a repulsion force between overlapping spheres is applied to reduce the overlap. After several repulsion steps the size of the container is adapted until a valid packing is obtained [5].

The CR approach can be substantially accelerated by using a Graphic Processing Unit (GPU) [25], which also enables the visualization using the same memory locations of the simulation results without the need of intermediate storage or copying of the results. This allows the CR simulation process to be visualized in real-time with specific structures like gaps and overlaps that can be used to judge the quality of the simulation [10].

2   Related Works

2.1   The Rasterization Approach

The visualization technique in [10] first rasterizes each sphere into pixels with information needed to determine the surface of gaps and overlaps. All such information will be stored in the global memory on the GPU to establish a GPU linked list [24]. After that, each pixel on the screen is assigned with a GPU thread. The GPU thread fetches information from the GPU linked list and sorts them locally. It can be observed that usually only information of spheres that are near to the observer is really needed and most information is stored but never used. Therefore, one can use ray tracing to avoid building a GPU linked list to store information for all pixels of all spheres in advance.

To render the overlaps based on rasterization, first all spheres are sorted based on their enter points with the frontmost sphere at beginning. Then it starts from the frontmost sphere and check all spheres based on their orders until an enter point lies between the enter and leave point of another sphere. This approach also supports a *periodic boundary* that allows a *tiling* of the sphere packing so that one unit cell can be used to simulate large spaces without influences from the boundary [7]. Under the Periodic Boundary Condition (PBC), spheres passing the boundary of the container will be duplicated on the opposite side and special cases need to be taken into consideration if the overlaps are out of the container.

The similar approach is applied to visualize the gaps formed by spheres under the PBC: Spheres (including those are duplicated due to the PBC) are sorted at first. If the first intersection point along the camera ray is the container boundary, the boundary will be rendered as gaps. If we first hit a sphere, we travel along the ray until we find a leave point which is *not in an intersection* between spheres. This point can then be rendered as the front side of the empty space.

2.2   Constructive Solid Geometry

Constructive Solid Geometry (CSG) is a technique to create a complex surface or object by using Boolean operators to combine simple objects. The overlaps of spheres can be interpreted as *intersections* of all spheres while the gaps can be seen as the difference between the container and all spheres. All CSG algorithms can be divided into two groups of approaches. The first approach calculates the boundary of the CSG first and then visualizes it with traditional 3D graphics tools [18–20]. Approaches in the second group are image-based techniques, where only an image of the CSG model from a given view is generated to avoid the complicated calculations of the complete boundary. The majority of methods of this kind are implemented by means of multi-pass which increases at least linearly with the scene's depth complexity [9,11,15,21]. Other approaches [14,16] in this group are based on ray tracing. Recently, Raza et al. [17] introduced a two-pass method based on GPU linked-list. Ulyanov et al. [22] extended the approach in [14] to a single pass two-stacks-based method which is capable of rendering CSG models consisting of more than a million CSG primitives. However, this approach can be further optimized for our problem to visualize overlaps and gaps formed by spheres.

3   Ray Tracing Based on Bounding Volume Hierarchies

In computer graphics, ray tracing is a widely used method that is composed of two elementary operations – acceleration structure traversal (ray traversal) and primitive intersection [2]. Many acceleration structures for ray tracing have been proposed in the literature. Bounding Volume Hierarchies (BVH), where a hierarchy of simple primitives (e.g. Axis Aligned Bounding Boxes (AABB's)) is used to exclude large parts of a scene in the ray traversal process, are among the most effective ones when used on a GPU [23]. In this paper, we use the Linear BVH (LBVH) in [12] for simplicity. However, the ray traversal approaches introduced in this paper can also be applied to many other types of BVHs such as those in [8,10,13] or the sophisticated LBVH construction algorithm in [4], where two separate steps (tree construction and bounding box calculations) are combined into a single kernel launch.

To visualize overlaps and gaps, the ray traversal approach needs to be adjusted since in comparison to the trivial ray traversal, which only needs to find the nearest intersection between the ray and the whole scene, the ray traversal for overlaps and gaps needs to record several intersection points until it finds the intersection point on the surface that is to be visualized.

3.1   Ray Tracing Framework

In this paper, we use stack-based ray tracing since it provides general good performance in comparison with stack-less variants in [1]. Binder et al. also provided a stack-less ray traversal method with backtracking in [6], which is faster than the stack-based approach applied in this paper. However, Bind's method requires specific adjustments for the BVH that slows down the BVH-construction process and is thus undesired for real-time applications. The approach in [3] is adapted in this paper and summarized in Algorithm 1. For each pixel, a ray that starts from the camera and passes the center of the pixel will be generated. And if the ray hits the container, it will traverse from the root of the BVH. The depth t of the pixel on the final image is initialized with the depth of the leave point (leave depth) on the container in respect of the ray as shown in function **Init** in list 2. Each time, two children of the current node is tested against the ray. If only one child node is hit by the ray, the current node is set to the child hit by the ray. If both children are hit by the ray, the current node is set to the child with the depth of the enter point (enter depth) closer to the origin of the ray, the other child is then put into the local stack. If both children are missed, the current node is set to the top-most node popped up from the stack. The whole process repeats itself until it needs to fetch the top-most node on the stack, but the stack has no node on it. Each time when the current node is set to a new node, it will first check whether we can skip the current node—if not, the depth of the current pixel t will be updated against the current node. In the end, the current pixel will be drawn based the value of t. In our implementation, the index of the sphere which provides the depth information will also be updated while updating t. However, this will not be shown in our algorithms for clarity. Definitions for functions **DrawPixelForOverlap** and **DrawPixelForGap** can be found in list 2. The following sections will explain the step **Update** t **based on node** for overlaps and gaps (summarized in Algorithms 2 and 3) in detail. For simplicity, the common data used for the following algorithms is shown in list 1.

Algorithm 1: Ray Tracing Framework.

---

1 `// initialization`
2 Init(t);
3 $node \leftarrow GetBvhRoot()$;
4 **while** *node != null* **do**
5     $c_0 \leftarrow leftChild(node)$;
6     $c_1 \leftarrow rightChild(node)$;
7     $t_0 \leftarrow EnterDepth(c_0, ray)$;
8     $t_1 \leftarrow EnterDepth(c_1, ray)$;
9   **if** $t_0 \neq null$ *and* $t_1 \neq null$ **then**
10   **if** t_1 *before* t_0 **then** Swap(c_0, c_1) ;
11        $node \leftarrow c_0$;
12   PushOnStack(c_1);
13   **else if** $t_0 \neq null$ **then** $node \leftarrow c_0$;
14   **else if** $t_1 \neq null$ **then** $node \leftarrow c_1$;
15   **else** $node \leftarrow PopOnStack()$; ;
16   **if** *node can be skipped* **then**
17   continue;
18   Update t based on node;
19 `// evoke one of the two functions depending on what we are drawing`
20 DrawPixelForOverlap(t) or DrawPixelForGap(t);

List 1: Common data for all algorithms.

Data: c_E : enter depth of the ray on the container.
Data: c_L : leave depth of the ray on the container.
Data: i: insert position for the current sphere.
Data: i_{start}: position of the first sphere among saved spheres.
Data: S: a list storing relevant spheres for the current ray.

3.2 Ray Traversal for Overlaps

It is to be noted that rendering overlaps without the PBC can be seen as rendering overlaps under the PBC where the enter depth of the container is set to 0 and the leave depth is set to positive infinity. Thus in the following parts of this section we only discuss the case under the PBC while the one without the PBC can be treated as a special case.

Pseudo code for how to update t against the current node is given in Algorithm 2. Each time an intersection is found between the ray and a sphere, the pair of enter point and leave point is inserted into a sorted local linked list on the depth value of the enter point in ascending order so that pairs with near enter points are at the beginning of the linked list. If an enter point lies between the enter and leave point of another pair, then it implies that this enter point lies on an overlap surface. However, it is not guaranteed that this point lies on the nearest overlap surface which needs to be visualized. The ray traversal needs to be conducted further until all spheres that need to be inspected lie behind the nearest overlap surface that is already found.

To accelerate the whole process, nodes behind the current overlap surface t (we also set the initial value of t to c_L so that nodes behind the container are skipped) are skipped. Besides, we only check spheres which intersect the container from the view of the ray (line 6). If there are already other spheres being checked before and saved in the local list, it will first try to find the overlap surface formed by the enter point t_E of the current sphere (saved to position i) by iterating from the frontmost enter points and checking whether t_E is in one of the saved spheres (lines 15–25). After this is done, if no overlap is found, we further check if an overlap can be formed by the enter point of a saved sphere that lies in the current sphere (lines 29–33), otherwise we can directly check the next sphere since the overlap formed by already saved spheres must lie behind the overlap surface we have just found.

3.3 Ray Traversal for Gaps

The ray traversal approach for gaps follows the similar principle as for overlaps but is more complicated. Since usually spheres lie near to the observer will be checked first on the BVH, which is fine if the nearest gap surface is not far away from the observer. However, in cases where the nearest gap surface for the current ray lies very far from the observer, lots of spheres need to be checked. To address this problem, we record the smallest possible depth values for all spheres

---

Algorithm 2: Update t based on node for rendering overlaps under the PBC.

---

1 **foreach** *sphere in the node* **do**
2   **if** *Ray hits the sphere* **then**
3        $t_E \leftarrow$ EnterDepth(ray, sphere);
4        $t_L \leftarrow$ LeaveDepth(ray, sphere);
5   **if** t_L *behind* c_E *and* t_E *before* c_L **then**
6           $S_i \leftarrow$ *sphere*;
7   **if** $i = 0$ **then**
8              $S[i].next = null$;
9              $i \leftarrow 1$;
10   continue;
11          $j \leftarrow i_{start}$;
12          $j_{last} \leftarrow null$;
13          $findOverlap \leftarrow false$;
14   **while** t_E *behind* $S[j].t_E$ **do**
15   // check overlap formed by S[i]
16   **if** $!findOverlap$ *and* t_E *before* $S[j].t_L$ **then**
17   **if** t_E *behind* c_E **then**
18                   $findOverlap \leftarrow true$;
19                   $t \leftarrow t_E$;
20   **else** return DrawContainerPixel() ;
21             $j_{last} \leftarrow j$;
22             $j \leftarrow S[j].next$;
23   **if** $j = null$ **then** break;
24          $S[i].next \leftarrow j$;
25   **if** $j_{last} = null$ **then** $i_{start} \leftarrow i$;
26   **else** $S[j_{last}].next \leftarrow i$;
27   // check overlap formed by S[j]
28   **if** $!findOverlap$ *and* $j! = null$ **then**
29   **if** $S[j].t_E$ *before* t_L *and* $S[j].t_E$ *before* t **then**
30   **if** $S[j].t_E$ *behind* c_E **then** $t \leftarrow S[j].t_E$;
31   **else** return DrawContainerPixel() ;
32          $i \leftarrow i + 1$;

---

to be checked and the gap surface candidate to enable a more sophisticated early termination of the traversal. Besides, since local memory is usually a bottleneck, we apply special memory management on the local linked list and merge pair records if they overlap with each other. The following parts of this section will explain our approach in detail.

The surface of a gap can be interpreted as the camera observes from outside of the container where the container becomes solid and the space occupied by spheres becomes empty, which implies we can only see the gaps as inner-structures of the container through holes where spheres passing through the boundary of the container. Based on this point of view, if the depth of the gap surface t for each ray hitting the container is never updated, the ray must hit the container before it hits any sphere, thus the container boundary will be rendered for the current pixel as shown in function **DrawPixelForGap**. And if all spheres that are not yet inspected lie behind the current overlap surface, then the current gap surface is the final gap surface that we are looking for, thus we can stop tracing. Besides, if we have never updated the gap surface and we find out that all unchecked spheres lie behind the closest enter point (noted as δ) between the ray and all checked spheres to the container boundary – which implies that all spheres lie behind the container and nothing except the container boundary can be seen, we can also cease our search now. To make the whole tracing process more efficient, we do not need to check the enter depths of all unchecked spheres one by one. Instead, we only need to check the enter depths of all unchecked nodes on the stack. However, if we only take the top-most node from the node stack as in most stack-based traversal algorithms, we have to iterate all nodes on the stack to get the frontmost enter depth of all unchecked nodes. To solve this problem, we introduce order to the stack: When we push a node to the stack, we maintain a linked list between these nodes based on the enter depths so that the first node has the frontmost enter depth. Each time when we need to take a node from the stack, we take the first node and we know all other nodes are behind it. Furthermore, if one node is taken from the stack, its position can still be reused for storing new nodes to reduce the local memory consumption. According to our observation, this optimization can make the trace process about 10 times faster for scenes tested by us.

Before the final gap surface is found, information of already checked spheres cannot be simply discarded, since unchecked spheres could connect the checked spheres to extend the gap surface found so far. However, this could consume a lot of local memory on the GPU. Therefore, if the information of a sphere needs to be saved, instead of saving it on a complete new position, we try to use it to update information of already saved spheres. For example in Fig. 1, spheres are checked in the order of $a \rightarrow b \rightarrow c \rightarrow d \rightarrow e \rightarrow f$. At first, a and b are stored in two separate positions after checking. Sphere c will be discarded after checking since it is completely in sphere b. Sphere d will not be stored on a separate position, instead, it will be used to update the enter depth of sphere a. Similarly, sphere e will be used to update the leave depth of sphere b. At last, after checking sphere f, the leave depth of sphere a will be updated to the leave depth of sphere e and the position occupied by sphere b can be used for other purposes. Now the camera can see from the enter point of sphere d to the leave point of e. This approach is summarized in lines 17–51 in Algorithm 3.

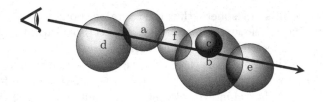

Fig. 1. Five spheres that in the end only need the storage for one sphere. Red areas are overlaps. (Color figure online)

It is to be noted that since only gaps in the container will be visualized, a gap surface found outside of the container will be ignored and treated as the ray passes through the container, which is described in list 2 for the function Update$t(t)$.

4   Results

We compared our new approaches with the rasterization methods in [10] on several test scenes with different size distributions (poli-disperse and mono-disperse spheres) and different scales (10 thousands, 1 million and 10 millions spheres) of spheres packings under the PBC. A test scene with one million poli-disperse spheres is shown in Figs. 2 and 3 shows the test scene with ten thousands mono-disperse spheres[1]. All performance measurements were conducted on an NVIDIA GeForce GTX 1080 graphics card with 8 GB video RAM. The PC is equipped with an Intel Core I7-6700K processor, 4.00 GHz CPU, 32 GB RAM, running Windows 10 (x64). If not indicated otherwise, the viewport resolution is set to 768 × 768 pixels.

First we compared their memory consumptions. The results are summarized in Table 1. The rasterization approaches need a large memory which is proportional to the number of rasterized pixels (16 bytes per pixel) of all spheres, regardless of whether those rasterized pixels will be occluded by other pixels. This implies that the memory requirement depends on the setting of our observation (position of the camera, field of view etc) and is hard to be predicted. Thus we have to pre-allocate a large bulk of memory for the GPU linked list whose size can only be approximated by experience. This can be seen in Table 1, where rasterization approaches need a lot of memory to render mono-disperse scenes due to that in each phase of CR, the total volume of all spheres are fixed and if the number of spheres is also given, the mono-disperse scenes will have the largest total area for spheres, which then generatesthe largest amount of pixels. On the

---

[1] The results are measured under non-zoomed view.

Algorithm 3: Update t based on node for rendering gaps under the PBC.

```
 1  foreach sphere in the node do
 2      if Ray hits the sphere then
 3          t_E ← EnterDepth(ray, sphere);
 4          t_L ← LeaveDepth(ray, sphere);
 5          if t_L behind t then
 6              if i = 0 then
 7                  S_i ← sphere;
 8                  S[i].next = null;
 9                  i ← 1;
10                  if t_E before t then
11                      // check if ray goes through the container
12                      if t_L behind c_L then                          // draw nothing
13                          ExitTracing();
14                  else δ ← t_E ;
15                  continue;
16              j ← i_start;
17              j_last ← null;
18              while t_E behind S[j].t_E do
19                  j_last ← j;
20                  j ← S[j].next;
21                  if j = null then break;
22              // S[k] and all spheres behind it are now behind t_E
23              k ← j;
24              // S[m] and all spheres before it are now before t_E
25              m ← j_last;
26              while j ≠ null and t_E before S[j].t_L do
27                  j_last ← j;
28                  j ← S[j].next;
29              // now S[j].t_E is behind t_L and S[j_last].t_E is before t_L
30              if m ≠ null and t_E before S[m].t_L then
31                  if j_last ≠ m then
32                      if S[j_last].t_L behind t_L then t_L ← S[j_last].t_L ;
33                      if t = S[m].t_L then Updatet(t_L) ;
34                      s[m].t_L ← t_L;
35                      s[m].next ← j;
36                  else if S[m].t_L before t_L then
37                      if t = S[m].t_L then Updatet(t_L) ;
38                      s[m].t_L ← t_L;
39                  else if j_last ≠ m then
40                      if S[j_last].t_L behind t_L then t_L ← S[j_last].t_L ;
41                      if t_E before t then Updatet(t_L) ;
42                      s[k].t_E ← t_E;
43                      s[k].t_L ← t_L;
44                      s[k].next ← j;
45                  else
46                      i ← i + 1;
47                      S[i] ← sphere;
48                      if m = null then
49                          if t_E before t then Updatet(t_L) ;
50                          else δ ← t_E ;
51                          i_start ← i;
52                      else S[m].next ← i ;
```

List 2: Function Definitions

```
 1  Function Init(t)
 2  |   t ← c_L;
 3  |   i ← 0;
 4  |   i_start ← null;
 5  |   S ← null;
 6  Function Updatet(t)
 7  |   if t behind c_L then    // ray goes through the container, draw nothing
 8  |   |   ExitTracing();
 9  |   t ← t;
10  Function DrawPixelForOverlap(t)
11  |   if t = c_L then  ExitTracing() ;
12  |   else  DrawOverlapSurfaceOn(t) ;
13  Function DrawPixelForGap(t)
14  |   if t = c_L then  DrawContainerSurface() ;
15  |   else  DrawGapSurfaceOn(t) ;
```
---

contrary, the memory consumption for Bounding Volume Hierarchy (BVH) can usually be calculated exactly in advance and lots of BVH building algorithms are memory efficient such as the LBVH used in this paper.

We also compared the runtime and summarized the results in Fig. 4. The BVH method is break down into four steps. Note that the time for **Duplicate Spheres** and **Build BVH** can be saved if spheres do not change their positions. For small number (10k) of poli-disperse spheres, the rasterization methods are a little faster than our new method due to the extra time used to build the BVH and duplicate spheres for PBC, which can be dropped out when spheres do not change their positions. In all other case, our method outperforms the rasterization method. Especially for mono-disperse scenes, the BVH method can be eight times faster. This improvement has two main contributions: One is that the GPU linked list is no longer needed and the LBVH can be built very fast; The other is that due to the BVH and our traversal approaches, lots of checks and calculations can be omitted at early stages.

5   Summary and Future Works

In this paper, we provide algorithms based on BVH to accelerate the visualization of gaps and overlaps for large and dynamic sphere packings. We implemented our methods with the widely used LBVH and compared our implementations with the rasterization methods. The results on the test scenes show that our methods can reduce an considerable amount of the memory consumption. In particular for mono-disperse spheres, the reduction can be 88% for ten thousands spheres. This also holds with respect to runtime and the improvement becomes more obvious when the amount of spheres increase. The exception is for ten thousands of poli-disperse spheres where our approaches are little slower than the rasterization

(a) Spheres

(b) Overlaps   (c) Gaps

Fig. 2. Spheres, overlaps and gaps for 1 million poli-disperse spheres.

methods due to extra steps that can be omitted when spheres no longer move. The reduced memory requirements and rendering time also enable us to render scene with more spheres (such as ten millions spheres in our test scenes.).

In the future, we would like to adjust our new approaches with more sophisticated ray traversal algorithms such those in [6] and [26] that have specific requirements on BVHs to achieve higher performance. Furthermore, we also plan to extend our methods for visualizing overlaps and gaps formed by particles in arbitrary forms.

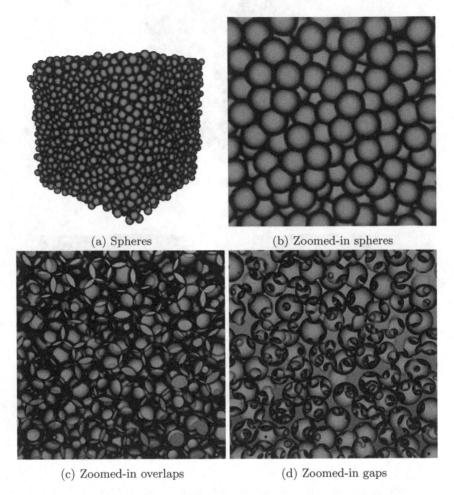

(a) Spheres

(b) Zoomed-in spheres

(c) Zoomed-in overlaps

(d) Zoomed-in gaps

Fig. 3. Ten thousands of mono-disperse spheres and their zoomed-in view for spheres, overlaps and gaps.

Table 1. Typical total memory consumption in Mb (N.A. for Not Available).

Number of spheres	10k		1M		10M	
Size distribution	Poly	Mono	Poly	Mono	Poly	Mono
Rasterization	108	318	167	665	N.A.	N.A.
BVH	37	37	112	112	868	868
BVH/Rasterization	34%	12%	67%	17%		

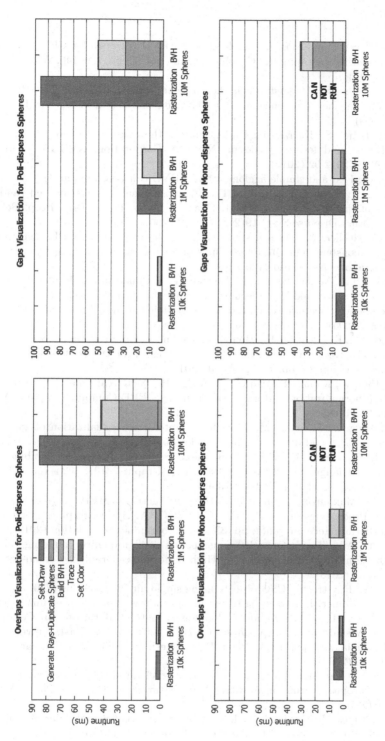

Fig. 4. Runtime comparison between the rasterization and BVH approaches.

References

1. Áfra, A.T., Szirmay-Kalos, L.: Stackless multi-BVH traversal for CPU, MIC and GPU ray tracing. In: Computer Graphics Forum, vol. 33, pp. 129–140. Wiley Online Library (2014)
2. Aila, T., Laine, S.: Understanding the efficiency of ray traversal on GPUs. In: Proceedings of the Conference on High Performance Graphics 2009, pp. 145–149. ACM (2009)
3. Aila, T., Laine, S., Karras, T.: Understanding the efficiency of ray traversal on GPUs-Kepler and fermi addendum. NVIDIA Corporation, NVIDIA Technical Report NVR-2012-02 (2012)
4. Apetrei, C.: Fast and simple agglomerative LBVH construction (2014)
5. Bezrukov, A., Bargieł, M., Stoyan, D.: Statistical analysis of simulated random packings of spheres. Part. Part. Syst. Charact.: Meas. Descr. Part. Prop. Behav. Powders Other Disperse Syst. **19**(2), 111–118 (2002)
6. Binder, N., Keller, A.: Efficient stackless hierarchy traversal on GPUs with backtracking in constant time. In: Proceedings of High Performance Graphics, pp. 41–50. Eurographics Association (2016)
7. Cheatham, T.I., Miller, J., Fox, T., Darden, T., Kollman, P.: Molecular dynamics simulations on solvated biomolecular systems: the particle mesh ewald method leads to stable trajectories of dna, rna, and proteins. J. Am. Chem. Soc. **117**(14), 4193–4194 (1995)
8. Domingues, L.R., Pedrini, H.: Bounding volume hierarchy optimization through agglomerative treelet restructuring. In: Proceedings of the 7th Conference on High-Performance Graphics, pp. 13–20. ACM (2015)
9. Goldfeather, J., Monar, S., Turk, G., Fuchs, H.: Near real-time CSG rendering using tree normalization and geometric pruning. IEEE Comput. Graphics Appl. **9**(3), 20–28 (1989)
10. Gu, F., Yang, Z., Kolonko, M., Grosch, T.: Interactive visualization of gaps and overlaps for large and dynamic sphere packings. In: VMV, pp. 103–110 (2017)
11. Hable, J., Rossignac, J.: Blister: GPU-based rendering of Boolean combinations of free-form triangulated shapes. ACM Trans. Graph. **24**(3), 1024–1031 (2005). https://doi.org/10.1145/1073204.1073306
12. Karras, T.: Maximizing parallelism in the construction of BVHs, octrees, and k-d trees. In: Proceedings of the Fourth ACM SIGGRAPH/Eurographics Conference on High-Performance Graphics, pp. 33–37. Eurographics Association (2012)
13. Karras, T., Aila, T.: Fast parallel construction of high-quality bounding volume hierarchies. In: Proceedings of the 5th High-Performance Graphics Conference, pp. 89–99. ACM (2013)
14. Kensler, A.: Ray tracing CSG objects using single hit intersections (2006)
15. Kirsch, F., Döllner, J.: Rendering techniques for hardware-accelerated image-based CSG. In: The 12-th International Conference in Central Europe on Computer Graphics, Visualization and Computer Vision 2004, WSCG 2004, University of West Bohemia, Campus Bory, Plzen-Bory, Czech Republic, 2–6 February 2004, pp. 221–228 (2004)
16. Lefebvre, S., Grand-Est, L.I.N.: IceSL: a GPU accelerated CSG modeler and slicer. In: 18th European Forum on Additive Manufacturing, AEFA 2013, Paris, France. Citeseer (2013)
17. Raza, J., Nunes, G.: Screen-space deformable meshes via CSG with per-pixel linked lists. GPU Pro 5: Advanced Rendering Techniques, p. 233 (2014)

18. van Rossen, S., Baranowski, M.: Real-time constructive solid geometry. In: Game Development Tools, vol. 79 (2011)
19. Schneider, P., Eberly, D.H.: Geometric Tools for Computer Graphics. Elsevier, Amsterdam (2002)
20. Segura, C., Stine, T., Yang, J.: Constructive solid geometry using BSP tree. In: Computer-Aided Design, pp. 24–681 (2013)
21. Stewart, N., Leach, G., John, S.: Linear-time CSG rendering of intersected convex objects. In: The 10-th International Conference in Central Europe on Computer Graphics, Visualization and Computer Vision 2002, WSCG 2002, University of West Bohemia, Campus Bory, Plzen-Bory, Czech Republic, 4–8 February 2002, pp. 437–444 (2002). http://wscg.zcu.cz/wscg2002/Papers_2002/B79.pdf
22. Ulyanov, D., Bogolepov, D., Turlapov, V.: Interactive vizualization of constructive solid geometry scenes on graphic processors. Program. Comput. Softw. **43**(4), 258–267 (2017). https://doi.org/10.1134/S0361768817040090
23. Vinkler, M., Havran, V., Bittner, J.: Bounding volume hierarchies versus Kd-trees on contemporary many-core architectures. In: Proceedings of the 30th Spring Conference on Computer Graphics, pp. 29–36. ACM (2014)
24. Yang, J.C., Hensley, J., Grün, H., Thibieroz, N.: Real-time concurrent linked list construction on the GPU. In: Computer Graphics Forum, vol. 29, pp. 1297–1304. Wiley Online Library (2010)
25. Yang, Z., Gu, F., Grosch, T., Kolonko, M.: Accelerated simulation of sphere packings using parallel hardware. In: Baum, M., Brenner, G., Grabowski, J., Hanschke, T., Hartmann, S., Schöbel, A. (eds.) SimScience 2017. CCIS, vol. 889, pp. 97–111. Springer, Cham (2018). https://doi.org/10.1007/978-3-319-96271-9_6
26. Ylitie, H., Karras, T., Laine, S.: Efficient incoherent ray traversal on GPUs through compressed wide BVHs. In: Proceedings of High Performance Graphics, p. 4. ACM (2017)

Simulation of Materials: Finite Element and Multiscale Methods

A Novel Approach to Multiscale MD/FE Simulations of Frictional Contacts

Henrik-Johannes Stromberg[1]($\boxtimes$), Nina Gunkelmann[2], and Armin Lohrengel[1]

[1] IMW of TU Clausthal, Robert-Koch-Straße 32, 38678 Clausthal-Zellerfled, Germany
stromberg@imw.tu-clausthal.de
[2] ITM of TU Clausthal, Adolph-Roemer-Straße 2A, 38678 Clausthal-Zellerfled, Germany

Abstract. In most applications, frictional contacts lead to a noticeable amount of wear, which influences the further frictional behavior. Thus, friction and wear have to be analyzed as a whole to gain powerful models. In such models the interactions of macroscopic and microscopic aspects have to be taken into account. Finite element (FE) simulations are the standard method to simulate macroscopic solid body mechanics. However, they are not suitable to represent microscopic behavior of bodies, especially abrasive friction depending on the roughness of the contact. In these applications, molecular dynamics (MD) simulations using explicit time integration schemes are a much better tool. The combination of both methods is an established approach for the solution of friction problems, which has been pursued by several authors. The usual way of linking both methods is to use MD domains for the boundary of contacting bodies modeled in FE instead of conventional contact elements. The interface between FE and MD domain is generally implemented by defining MD particles and FE nodes as coincident. With this approach, every time step of the MD simulation requires solving a linear equation system for the whole FE modeled solid. The computational cost of solving a sparse linear equation system is superlinearly dependent on its degrees of freedom. Furthermore, MD simulations use explicit time integration, which requires very small time steps to assure stability. Thus, it is disadvantageous to apply the current coupling method to large geometries, since large linear equation systems would have to be solved very often. In consequence, a different approach is required to apply multiscale MD/FE methods to complex geometries.

This paper introduces a novel approach to integrate multiscale capabilities into FE that allows solving large models at reasonable computational cost. The proposed approach integrates MD coupling into FE contact elements characterized by a nonlinear, history dependent friction law, which is trained with MD simulations. The roughness profile of sliding surfaces is modeled with elasto-plastic spherical caps serving as a mesoscopic level. The Hertzian contact between two spherical caps is handled at the microscopic level using an improved variant of the conventional node particle coincidence technique.

Keywords: Multiscale simulation · Finite elements · Molecular dynamics simulation

N. Gunkelmann and M. Baum (Eds.): SimScience 2019, CCIS 1199, pp. 153–167, 2020.
https://doi.org/10.1007/978-3-030-45718-1_10

1 Introduction

1.1 Microscopic Molecular Dynamics

Molecular dynamics simulations of frictional contacts model tribological processes on an atomistic scale and allow obtaining insights into both the process dynamics and the detailed structure of the final state [1]. For more than 20 years, the problem of contact between two bodies can be studied by atomistic simulations [2]. At the nanoscale, surface forces dominate the tribological behavior of the system and friction coefficients vary depending on the system conditions [3]. Luan and Robbins [4, 5] showed that nanoscale contacts are determined by processes at the atomistic scale including dislocation nucleation and plastic deformation of asperities. Solhjoo et al. [6] studied tribological contacts of different roughness using molecular dynamics simulations and compared their results with continuum theories. The authors found significant deviations in the elastic moduli. Since in situ experiments of friction phenomena at the atomistic scale are very rare, molecular dynamics simulations can help understanding the underlying physical mechanisms. Therefore, these phenomena need to be studied by atomistic modeling techniques because the tribological properties of frictional contacts are highly influenced by adhesion, bond formation and bond breaking at the nanoscale.

In recent years, the number of MD simulations of tribological systems has increased drastically. One reason is that modern force fields are significantly improved to reproduce the material behavior of complex molecules [7, 8]. Secondly, recent advances in high performance computing architectures have enabled larger molecules and timescales.

However, in MD, only small nanometric scales are available. Therefore, many researchers focus on the sliding of a single asperity where the interface is only a few nanometers in size [9–11]. Often, also the scratching of a surface is studied with MD to investigate nanoscale deformation and wear mechanisms [12, 13].

Since the invention of the atomic force microscope (AFM) in 1987 [14], single-asperity frictional contacts can be directly measured allowing comparisons with MD data. The atomic interaction at the interface between tip and surface cannot be captured by experiments and therefore, MD simulations usually complement AFM measurements [2, 15, 16].

The single-asperity problem, however, is mostly relevant for applications using nanometric devices and fails to predict large-scale tribological applications. In a paper by Blau [17], it was shown that friction is highly scale dependent making it necessary to model the entire tribological system instead of single asperities. Thus, to capture the behavior of macroscopic tribological systems, continuum models must be taken into account.

1.2 State of the Art of Multiscale Simulations in Tribology

As stated above, MD simulations are not suitable to model the behavior of large volumes given that the resulting number of particles and in consequence the number of degrees of freedom exceeds the scale for practical computation. In consequence multiscale methods combing MD and FE were developed. Ramisetti et al. provide an overview of the techniques currently available [18]. The common approach is to model solids as a continuum with finite elements and use discrete particles in an MD model for near-surface areas. This method requires an interface between both regions. The interface is then modeled by MD particles, which are also FE nodes. This method is often called Quasi-Continuum method (QC) [19]. However, for these methods, the implemented interfaces vary in existence and density of an intermediate region (see Fig. 1), as implemented in the Bridging Domain method [20].

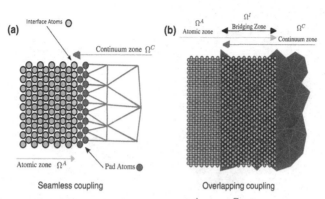

Fig. 1. Coupling methods for MD/FE interface [18]. Ω^A and Ω^C refer to atomistic and continuum zone, while the overlapping or bridging zone Ω^I has been introduced in the Briding Domain Method [20].

Ramisetti et al. [18] describe a seamless coupling where all particles of the interface surface are coincident with nodes in contest to coupling with a bridging zone where the FE mesh spans over a larger partition of the MD domain with coincident nodes distributed over the bridging zone. It is obvious that the seamless coupling method leads to a mesh containing elements with bad aspect ratios at the interface. The discrete maximum principle [21] enforces the sum of inner angels of two triangular elements, facing a common edge not to exceed π in order to guarantee convergence. The seamless coupling approach, however, generates elements, which fail this criterion. Accordingly, convergence cannot be guaranteed.

In another work [22], Sun and Cheng also use an intermediate method using all particles in the overlapping region as FE nodes, but leaving voids in between (see Fig. 2). This approach also generates elements, which are not compliant with the discrete maximum principle. In all cases, the displacement field calculated in FE is interpolated between FE nodes using the FE shape functions used by the applied finite elements. In conclusion, the number of particles between nodes which are given interpolated displacements is the key parameter which can be altered in order to improve element aspect ratio. Selecting a large parameter however, will result in growing interpolation errors, especially when used with linear element shape functions.

Note that there exist many methods for molecular statics atomistic-continuum coupling. The energy-based quasi-continuum method [19], the coupling of length scales method [23] and the bridging scale method [20] may be mentioned. However, frictional contacts are dynamic problems involving time-dependent boundary conditions, displacements, atomic forces and continuous tracking. In the atomistic region, the trajectory of the atoms needs to be studied, while in the FE region, mean displacement, velocities and temperature fields are approximated. Nonequilibrium processes require the modeling of wave propagation and heat transfer and are therefore challenging. A common method extents the QC-method to local-harmonic approximations at constant temperature [24]. In this work, we will use the Atoms-to-Continuum approach by Wagner et al. [28] where thermal fluctuations in the MD domain are coupled to an energy equation in the continuum.

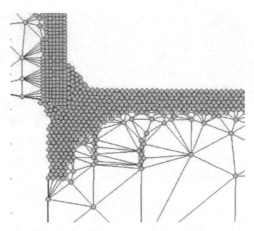

Fig. 2. Intermediate method using all particles in the overlapping region as FE nodes, but leaving voids in between [22]

2   How Can Multiscale MD/FE Become More Potent?

2.1   Computational Cost of Current Approach

When MD and FE are coupled, both algorithms have to be executed sequentially for every time integration step. In this situation, the time step size is set to the stability radius. Compared to usual time step sizes in implicit nonlinear FE, very small time steps have to be used.

The whole calculation is performed in a loop over all time steps t_n. In each timestep, new positions, velocities and accelerations of all particles in the MD domain are computed. Furthermore, deflection and reaction forces of the continuum-modeled body are calculated. The whole process is summarized in Fig. 3.

The most costly step in terms of computation time in each iteration is solving the linear equation system (LES) generated from FE. In order to clarify this point, it has to be noted that all available solvers for sparse linear equation systems have a superlinear cost function. This means, that solving a problem with twice as many degrees of freedom, will cost more than twice as much computation time, regardless of the circumstances. The number of degrees of freedom typically scales with volume. In consequence, an algorithm, which solves multiple smaller systems, is substantially faster than an algorithm that solves one big system. Therefore, the maximum mesh size for affordable computation time is limited to rather small problems for the known algorithm. Furthermore, the algorithm cannot divide a problem in multiple contact zones in order to reduce computation cost by utilizing their independency from each other. As forces between all atoms within a cutoff distance have to be calculated in the computation, time has a quadratic dependency on the number of MD atoms. Since contacts are generally possible on all surfaces, the number of atoms is linear with respect to the model surface. Taking all stated dependencies into account, the global computation time can be approximated by Eq. (1) where A is the model surface area, V the model volume and n_t the number of time steps.

$$t = n_t \cdot \left(\mathcal{O}\left(A^2\right) + \mathcal{O}\left(V^k\right)\right) k \in \{1 \ldots 2\} \cap \mathbb{R} \tag{1}$$

The notation $\mathcal{O}$ describes the asymptotic behavior of computational cost. The constant k depends on the solving algorithm for linear equations systems, which is used. In the worst-case scenarios, $k = 2$ can be assumed.

2.2 Target Characteristics for Full Part Simulation

The general idea of coupling finite element and molecular dynamics methods is very promising for a variety of practical engineering problems concerning contact behavior. All conventional approaches to predict wear resistance of machine parts are based on tests with the target geometry or have low reliability. Furthermore, they cannot predict the frictional behavior in different stages of wear. This gap in computability of machine parts could potentially be filled using a multiscale MD/FE approach, which solves frictional contacts and their wear as a joined problem. However, the application of multiscale methods for problems in real-world scale requires benign numerical characteristics of the used method. They are based on the basic need of computing problems without super computers and in short time. This section describes such target characteristics.

For many practical problems, a three dimensional geometry model is required. In order to ensure small errors in FE, fine meshes with a magnitude of 10^5 Elements are used. Given that most mechanical problems include bending, e.g. bending of gear teeth, solid elements with quadratic shape functions have to be used. Typical 20 knot brick elements have 60 degrees of freedom. In consequence, it can be stated that $n_{FE} \approx 10^6$ which is vastly more than used in published work with MD/FE coupling [1, 18]. As wear is mostly caused by kinematically complex interactions of surfaces such as eccentric movement of spline hub connections, the behavior of multiple contact points has to be considered. In consequence, all surfaces which may touch have to be equipped with MD active zones leading to a very high number of MD particles. Published work focuses on problems with one contact zone. In order to reduce computation time and improve possibilities for parallelization, which is required to achieve good performance on modern computer architectures, it would be highly advantageous to reduce data dependencies between both scales. Furthermore, the model should decouple time step sizes of both levels.

When an algorithm, which meets the requirements stated above, is developed and implemented, it should allow wear simulations, which are independent of tests with

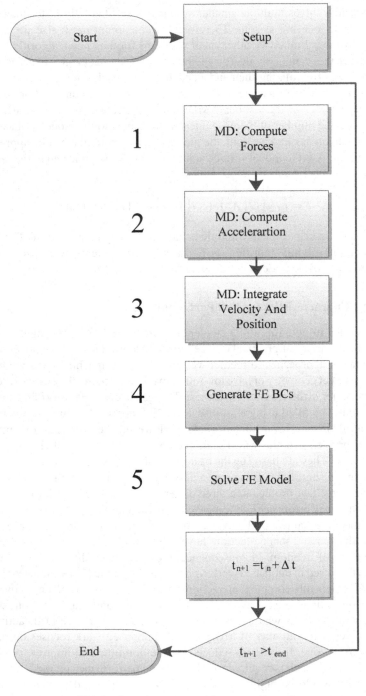

Fig. 3. Complete process of current approach

actual part geometry reducing the amount of expensive experiments, and improving the understanding of part failure.

3 Proposed Approach

3.1 Global Setup

The basic idea of the proposed approach is to decompose the numeric problem by introducing an intermediate level into the simulation setup. The global setup of the calculation method is shown in Fig. 4. At the macroscopic level, a conventional FE model is used. Is utilizes modified contact elements with internal variables q (local material memory behavior) as known from nonlinear material models. The contact element itself performs the contact detection and data handling. Each contact element uses a 2-dimensional representation of its surface by sphere caps as an intermediate level to calculate the frictional force F_F from normal force F_N, sliding velocity v and internal variables. The sphere caps model is derived from measured surface profiles based on the method of Greenwood and Williamson [25]. It is applied for example in the simulation of thrust cones by Heß [26].

The sphere caps are modeled as rods with elastoplastic behavior and one degree of freedom in the direction of the contact normal. This setup allows the surface to smooth out in contact as it is observed in experiments as run-in behavior.

The contact between individual sphere caps is handled in the micro-level, which is responsible for computing the molecular contact in order to calculate local friction coefficients μ and the linear wear coefficient. The micro-level model uses a classical QC setup. It is provided with radii of the interacting caps, base material, normal force F_N, contact velocity v, angel of contact normal φ and the partition of inner variables used for the molecular dynamics setup q_{MD}. With these input parameters, a simulation is set up and solved. The internal variable q_{MD} contains information about the composition of the MD modeled layer and MD boundary conditions. It is necessary to carry this local information through the global time steps to account for effects such as wear of coatings. By storing it as an internal variable, it is available to the FE post processor fur further evaluation. Particles are considered part of one of the two contact partners or the quasi-liquid phase in between [27]. In consequence, q_{MD} contains the composition of all the domains. When a particle detaches from a contact partner, it is moved to the quasi-liquid phase domain and a new particle is generated in the transition layer according to the used MD boundary conditions. This tool allows simulating effects such as the depletion of coatings by lowering the fraction of coating particles generated and generating more base-material particles. When a particle reaches the end of the simulated domain to the right or left, it is removed from the semi-fluid domain.

The simulation results of the micro-level are then used in the second level to compute the total frictional force F_F. From v_{lin}, the wear induced shortening of each cap is calculated.

The transition between the three phases of wear can be detected by the changing composition of the semi-fluid phase such as rising fraction of base material. Due to the massive geometry changes in the end phase of wear, it is not possible to simulate

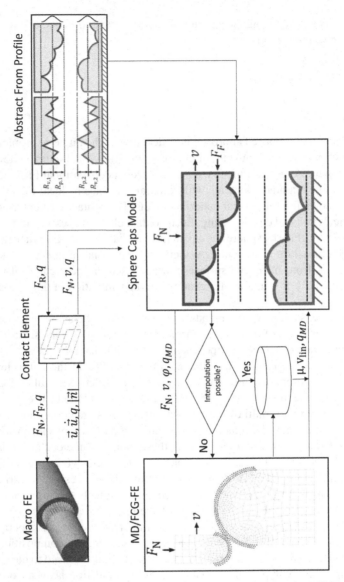

Fig. 4. Global calculation setup (sphere caps visualization from [26])

this process with the proposed method. The failure criterion for a machine part is its transition of a significant number of elements to the quasi-liquid phase.

The proposed procedure will speed up the simulation process but still take long computation time for the micro-level. In order to reduce the required computation time at the micro-level, an automatic problem adapted algorithm is introduced. Note that a comparison of the approach presented here with standard models will be investigated in the course of our further research. The focus of this paper is the optimization of the micro-level simulations in Sect. 3.3.

3.2   Micro-level Acceleration

The crucial factor in terms of computation time for the proposed algorithm is the micro-level. Therefore, the focus of further optimization is put on accelerating it. The two obvious aspects, which should be subjected to optimization, are the number of executions of the micro-level and the number of operations required for each execution.

In order to execute micro-level simulations less frequent, the result of all micro-level simulations is saved to a database. Before a simulation is performed, the database is checked for results with input parameters near the new point. If an interpolation of existing data is possible, a new simulation is not necessary. However, a criterion for interpolation errors has to be established. Whenever a simulation is conducted, its results are compared to an interpolation solution. By doing so, an array of estimated interpolation errors for various regions of the database can be formed allowing to follow specified error criterions. This procedure lets the system learn relevant situations in the micro-level for a specific class of problem. In conclusion, the system is problem-adaptive and trains solving problems. It thereby becomes faster over time.

Solving the micro-level simulation requires to solve two linear equation systems, one for each sphere cap of the contact for every step in the MD time integration. This process is responsible for the bulk arithmetic operations on the micro-level and therefore should be subject to further optimization.

There are several direct solvers for linear equation systems, which require a first computing intensive step, which is independent of the right hand side and therefore can be computed without knowledge of the right hand side. This step can be performed at compile time for all sphere cap sizes. With this setup, mesh generation and matrix preparation are out of consideration for the complexity of the algorithm. The most adequate options for such algorithms are using a factorization or computing the inverse of the matrix. As done at compile time, the significantly higher computation time of inverting a matrix compared to a factorization is not relevant. Due to Newton's first law, the matrix is symmetric. This can be utilized by performing a Cholesky decomposition as factorization. The decisive factor is the resulting performance in solving the preprocessed equation system. One of the key factors here is the so-called fill-in, which describes how many nonzero elements the processed matrix gains compared to the original matrix. A massive gain on non-zero elements is disadvantageous for the operation of solving the linear equation system as it enlarges the number of elements to be processed and the memory consumption hence only nonzero elements are stored in the used sparse matrix format.

This new method was integrated into the source code of the Atoms-to-Continuum package (AtC) in the open source molecular dynamics code LAMMPS. The user package includes FE definitions, operators, interpolators and time integrators to couple atomistic to continuum methods. In this way, classical molecular dynamics simulation is used in regions where the dynamics of the system is too complex to describe it with continuum laws, e.g. in regions containing a larger number of defects, dislocations and phase boundaries. To this end, a domain is discretized with a finite element mesh. An internal part of the domain is filled by a set of atoms. The normal vector of the boundary between FE and MD domain is oriented into the MD region. To derive a coupled MD/FE equation system, it is assumed that the total energy of the system can be divided into two parts corresponding to the energy of both subdomains. The energy of the MD domain is given by the sum of the kinetic and potential energy of the atoms. The energy of the FE system corresponds to the sum of strain energy, kinetic energy of the FE region and the thermal energy integrated over the FE region. In the AtC package, only the thermal energy contributes to the energy of the FE system because AtC focuses on heat transfer applications. To incorporate the effect of the temperature of the continuum region on the MD region, a drag force acts on the atoms. By enforcing conservation of energy for the total system, the equation for the nodal coefficients is derived. Details of the method are given in ref. [28].

As the focus of the implementation lies on maximizing the speed of solving the system with the decomposition, a data structure that fulfils this purpose was selected. The factorization itself is performed in place in a C++ map, which is used as a sparse matrix. Afterwards, this structure is converted into two vectors representing the same sparse matrix and its transposition with preordered elements. Each vector element contains the matrix locations m and n as well as the value. The vectors nzi and nzit are ordered in the way array elements are accessed for forward substitution and back substitution respectively. The vector nzit contains the transposed matrix. The code for solving the equation system for a specific right hand side is shown in Fig. 5.

```
1     void CholDecomp::solve(std::vector<double>* b, std::vector<double>* x)
2     {
3         std::vector<double> y;
4         y.resize(dof);
5         x->resize(dof);
6         double lhs = 0;
7         for (std::vector<arritem>::iterator it = nzi.begin(); it != nzi.end(); ++it)
8         {
9             if (it->m != it->n)
10            {
11                lhs += it->val*y[it->n];
12            }
13            else
14            {
15                y.at(it->m) = (b->at(it->m)-lhs)/it->val;
16                lhs = 0;
17            }
18        }
19
20
21        for (std::vector<arritem>::reverse_iterator it = nzit.rbegin(); it !=
          nzit.rend(); ++it)
22        {
23            if (it->m != it->n)
24            {
25                lhs += it->val*x->at(it->n);
26            }
27            else
28            {
29                x->at(it->m) = (y.at(it->m) - lhs) / it->val;
30                lhs = 0;
31            }
32        }
33    }
```

Fig. 5. Source code for solving

The proposed method leads to substantial gain in computation speed of the AtC package as is shown in Fig. 6. Depending on the test case, a performance gain between factor 53 and 95 was measured.

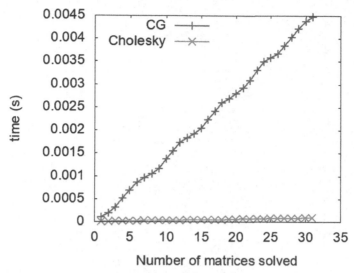

Fig. 6. Comparison of LES solvers in AtC

3.3   Test Implementation of Wear Using the Atoms-to-Continuum Package

To demonstrate the effect of micro-level acceleration, we study a system of the contact between two particles where the MD region includes the contact region (see Fig. 7). The MD system is given by fcc Cu. Interatomic forces are given by the EAM potential by Foiles et al. [29]. The computational domain comprises a spherical region of radius 20 Å and the FE particles have a radius of 100 Å. The MD domain comprises a radius of 1.5 nm. The FE region consists of a hexahedral mesh with a mesh size of 2000 hexahedral elements with linear shape functions and 8 nodes each. The geometry and meshes are generated with a utility called blockMesh within the open-source software OpenFOAM [30]. The mesh has 2662 nodes. We use free boundary conditions in all directions. We minimize the energy by a conjugate gradient algorithm with an energy tolerance of 10^{-25} in LAMMPS (with the meaning that the energy change between successive iterations divided by the energy magnitude is less than or equal to the tolerance) and a force convergence value of 10^{-25} eV/Å. The relative velocity of the sliding sphere is set to 0.1 m/s. The system evolves for 50 ps under NVE conditions and estimates of free energy density, displacement vectors and gradients are calculated.

In the post processing, the material fraction, which is transferred from the caps into the quasi-liquid phase has to be calculated. This is achieved by using the idea of a cutoff distance after which forces between particles are neglected. The method is commonly used in LAMMPS to limit computation time and the influence of floating point arithmetic errors. Here it is used to recognize independent particles. A particle or a group of particles is seen as independent, if it is not connected to particles outside the group by a distance below cutoff. This technique is used to identify all particle groups in the post simulation configuration. The largest group, which was part of the first spherical cap in the pre simulation setup, is defined to represent the first spherical cap. The same procedure

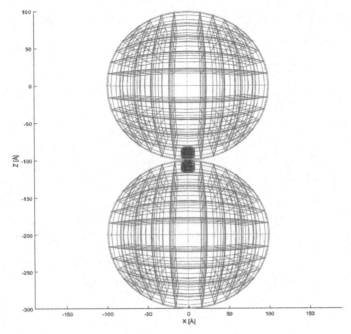

Fig. 7. Micro-level implementation setup

is used for the second sphere cap. All other groups are then added to the quasi-liquid domain. From mass loss, wear is calculated by computing the equivalent loss in radius if a sphere shape is maintained. It has to be investigated, whether a correction factor for flattening of the caps is required to improve accuracy.

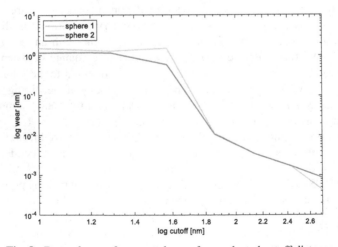

Fig. 8. Dependency of computed wear from selected cutoff distance

As it is obvious, the cutoff distance has a high influence on the computed wear. Figure 8 shows the wear-cutoff dependency for the described computing example. In the future, a plausible calculation method for the post processing cutoff distance has to be found.

4 Conclusion and Future Investigations

The proposed approach has several traits, which improve its applicability to large and complex geometries:

- Linear computation cost for number of contact zones
- Independency of time step sizes at micro- and macro-level
- Size of MD coupled mesh is independent of macro geometry
- Capability to account for mesoscopic and microscopic run-in and wear
- Microscopic condition can be evaluated by FE post processor

Our results show that by optimizing the solvers used for MD/FE coupling at the micro-level, the computation speed of multiscale simulations can be significantly accelerated. The progress, which has been made in improving the computation speed of the Atoms to Continuum package in LAMMPS will be made available to the public by integrating it into the main branch of AtC.

Future investigations will deal with the following points:

In order to explore the potential of the method and compare its results against state of the art, it should be implemented into a scientific FE code. In order to be able to validate the approach, a machine part should be modeled which has relevant wear failure and can be simulated under the restrictions of the method. One possible candidate would be a dry lubricated spline hub connection.

In the software development process, special attention should be payed to the used interpolation criteria. Experiments with different criteria may provide significant improvement to the algorithms performance. As stated before, different preprocessing techniques should also be tested.

Lastly, further improvement of the method should be considered. As it is currently proposed, the method is not capable of modeling the shearing of spherical caps, which will occur for more rough surfaces. Sheared-off material would have to be introduced as free particles in the mesoscopic level simulation. This would greatly increase the complexity of algorithm development and calculation.

References

1. Luan, B.Q., Hyun, S., Molinari, J.F., Bernstein, N., Robbins, M.O.: Multiscale modeling of two-dimensional contacts. Phy. Rev. E **74**, 046710 (2006)
2. Bhushan, B., Israelachvili, J.N., Landman, U.: Nanotribology: friction, wear and lubrication at the atomic scale. Nature **374**, 607 (1995)

3. Kim, H.J., Kim, D.E.: Nano-scale friction: a review. Int. J. Precis. Eng. Manuf. **10**, 141–151 (2009). https://doi.org/10.1007/s12541-009-0039-7
4. Luan, B., Robbins, M.O.: The breakdown of continuum models for mechanical contacts. Nature **435**, 929 (2005)
5. Luan, B., Robbins, M.O.: Hybrid atomistic/continuum study of contact and friction between rough solids. Tribol. Lett. **36**, 1–16 (2009). https://doi.org/10.1007/s11249-009-9453-3
6. Solhjoo, S., Vakis, A.I.: Continuum mechanics at the atomic scale: Insights into non-adhesive contacts using molecular dynamics simulations. J. Appl. Phys. **120**, 215102 (2016)
7. Siu, S.W.I., Pluhackova, K., Böckmann, R.A.: Optimization of the OPLS-AA force field for long hydrocarbons. J. Chem. Theory Comput. **8**(4), 1459–1470 (2012)
8. Ewen, J.P., Gattinoni, C., Thakkar, F.M., Morgan, N., Spikes, H., Dini, D.: A comparison of classical force-fields for molecular dynamics simulations of lubricants. Materials **9**(8), 651 (2016)
9. Cha, P., Srolovitz, D.J., Vanderlick, T.K.: Molecular dynamics simulation of single asperity contact. Acta Mater. **52**, 3983–3996 (2004)
10. Jeng, Y., Su, C., Lay, Y.: An investigation of nanoscale tribological characteristics under different interaction forces. Appl. Surf. Sci. **253**, 6754–6761 (2007)
11. Yang, J., Komvopoulos, K.: A molecular dynamics analysis of surface interference and tip shape and size effects on atomic-scale friction. J. Tribol. **127**, 513–521 (2001)
12. Tran, A.S., Fang, T.H., Tsai, L.R., Chen, C.H.: Friction and scratch characteristics of textured and rough surfaces using the quasi-continuum method. J. Phys. Chem. Solids **126**, 180–188 (2019)
13. Gunkelmann, N., Alhafez, I.A., Steinberger, D., Urbassek, H.M., Sandfeld, S.: Nanoscratching of iron: a novel approach to characterize dislocation microstructures. Comput. Mater. Sci. **135**, 181 (2017)
14. Mate, C.M., McClelland, G.M., Erlandsson, R., Chiang, S.: Atomic-scale friction of a tungsten tip on a graphite surface. Phys. Rev. Lett. **59**, 1942 (1987)
15. Braun, O.M., Naumovets, A.G.: Nanotribology: microscopic mechanisms of friction. Surf. Sci. Rep. **60**, 79 (2006)
16. Gnecco, E., Bennewitz, R., Gyalog, T., Meyer, E.: Friction experiments on the nanometre scale. J. Phys. Condens. Matter **13**, 619–642 (2001)
17. Blau, P.J.: Scale effects in steady-state friction. Tribol. Trans. **34**(3), 335–342 (1991)
18. Ramisetti, S.B., Anciaux, G., Molinari, J.-F.: MD/FE multiscale modeling of contact. In: Gnecco, E., Meyer, E. (eds.) Fundamentals of Friction and Wear on the Nanoscale. NT, pp. 289–312. Springer, Cham (2015). https://doi.org/10.1007/978-3-319-10560-4_14
19. Tadmor, E.B., Ortiz, M., Phillips, R.: Quasicontinuum analysis of defects in solids. Phil. Mag. A **73**, 1529 (1996)
20. Xiao, S.P., Belytschko, T.: A bridging domain method for coupling continua with molecular dynamics. Comput. Methods Appl. Mech. Eng. **193**, 17–20 (2004)
21. Ciarlet, P.G., Raviart, P.A.: Maximum principle and uniform convergence for the finite element method. Comput. Methods Appl. Mech. Eng. **2**, 17–31 (1973)
22. Sun, X., Cheng, K.: Multi-scale simulation of the nano-metric cutting process. Int. J. Adv. Manuf. Technol. **47**, 891–901 (2010). https://doi.org/10.1007/s00170-009-2125-5
23. Abraham, F.F., Broughton, J.Q., Bernstein, N., Kaxiras, E.: Spanning the length scales in dynamic simulation. Comput. Phys. **12**, 538 (1998)
24. Dupuy, L.M., Tadmor, E.B., Miller, R.E., Phillips, R.: Finite-temperature quasicontinuum: molecular dynamics without all the atoms. Phys. Rev. Lett. **95**(6), 060202 (2005)
25. Greenwood, J.A., Williamson, J.B.P.: Contact of nominally flat surfaces, In: Proceedings of the Royal Society of London Series A, Mathematical and Physical Sciences, pp. 295–1442 (1966)

26. Heß, M.: Einsatz von Druckkämmen zur Effizienzsteigerung von schrägverzahnten Getrieben, doctoral thesis at TU Clausthal (2018). ISBN 978-3-86948-624-6
27. Bucher, F., Dmitriev, A.I., et al.: Multiscale simulation of dry friction in wheel/rail contact. Wear **261**, 874–884 (2006)
28. Wagner, G.J., Jones, R.E., Templeton, J.A., Parks, M.L.: An atomistic-to-continuum coupling method for heat transfer in solids. Comput. Methods Appl. Mech. Eng. **197**, 3351–3365 (2008)
29. Foiles, S.M., Baskes, M.I., Daw, M.S.: Embedded-atom-method functions for the fcc metals Cu, Ag, Au, Ni, Pd, Pt, and their alloys. Phys. Rev. B **33**, 7983–7991 (1986)
30. Weller, H.G., Tabor, G., Jasak, H., Fureby, C.: A tensorial approach to computational continuum mechanics using object-oriented techniques. Comput. Phys. **12**, 6 (1998)

Numerical Investigation of Tri-Axial Braid Composite Structures as Crush Specimens Using the VPS-Solver

Sven Hennemann[1]([✉]), Volker Hohm[2], Peter Horst[3], Lasse Twardy[1], and Philip Zimmermann[1]

[1] Volkswagen Group, Brand Volkswagen Pkw, 38440 Wolfsburg, Germany
[2] Volkswagen Group, Group Innovation, 38440 Wolfsburg, Germany
[3] Institute of Aircraft Design and Lightweight Structures, 38108 Brunswick, Germany

Abstract. This paper describes a new modelling approach for numerical investigations on the crushing behavior of tri-axial braid composite structures using the Virtual Performance Solution (VPS)-solver (also known as PamCrash; [1]) by way of its application to automotive related specimens.

In current VPS software, the crushing mechanism is modelled by a special crushing ply (PLY100), which is capable to model the mentioned mechanism phenomenologically but not able to behave as a splaying material which can possibly interact with other materials or itself [4]. With the presented novel approach, the visual structural behavior of a fiber reinforced plastic (FRP) sample and the shape of the corresponding load-displacement-curve can be simulated quiet accurately, especially for plane and oblique impactors. It is based on shell elements which are arranged in stacked order and implies a developed modelling strategy. In addition, a cohesive interface is defined to join the shell layers with each other. This interface acts as a matrix phase mixed with characteristics of braid yarns and contains a model for delamination. For experimental validation, a small extract was resorted of a large experimental study using a drop-weight-tower [8].

1 Introduction

Lightweight design with FRP materials is still an important and challenging mission in the development of current and future vehicles. The weight reduction generally increases vehicle range and dynamics, hence generally improving the driving experiences with electric vehicles. The usage of FRP composites significantly improves the efficiency of energy absorbing structures, such as crush specimens. Textile structures such as braids are particularly suitable as crushing structures due to the implemented fiber ondulations, which greatly improve energy absorption. The failure mechanism of *crushing* combines elements of many different single failure mechanisms inside the composite on different scales. While the combination of failure mechanisms increases the energy absorption greatly, it also illustrates the simulative challenges for various model and simulation techniques to yield a high accuracy.

© Springer Nature Switzerland AG 2020
N. Gunkelmann and M. Baum (Eds.): SimScience 2019, CCIS 1199, pp. 168–178, 2020.
https://doi.org/10.1007/978-3-030-45718-1_11

While there is extensive experience in the industry regarding experimental treatment of FRPs, vehicle design engineers seeks accurate tools for their prediction in numerical simulation, especially for crash simulation and *crushing* conditions. Commercially available software and literature show limitations regarding the accurate prediction of FRP *crushing* regarding the requirements for automotive engineers [6, 7]. In an earlier paper [8] the first two approaches (see Fig. 1, number one and two) concerning this modelling approach have been presented.

The work done offers a new phenomenological modelling approach, which has been developed in link with master thesis [10, 11] and is presented in detail in this paper. Details regarding FRP-material behavior and different failure criteria are presented there, too. The presented Multi Shell Layer (MSL)-Strategy is based on shell elements which are arranged in stacked order, i.e. the laminate stack is subdivided into several layers and each of those are modelled by one separate shell element surface, and implies a developed modelling strategy. In advance a specification sheet testing was conducted.

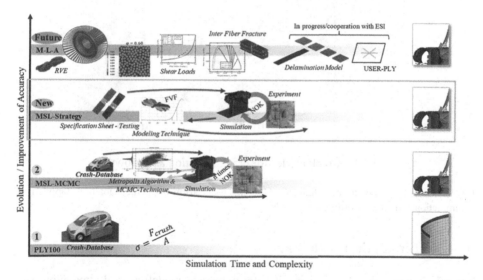

Fig. 1. Modelling approaches and evolution (MCMC method: Markov Chain Monte Carlo Method| M-L-A: Multi level analysis| FVF: fiber volume fraction)

2 Material Behavior and Failure of FRP

Composites have complex in plane and interlaminar ply failure mechanisms on different scales that may occur independently or even coupled. Those can be subdivided into three groups: fiber failure, matrix failure and delamination. The failure mode *crushing* contains all mentioned mechanisms. Due to the high number local failures and therefore high total energy absorption, it well suited for safety related applications.

Figure 2 shows experimentally observed damage and failure phenomena [1, 9]. Crack growth result in delamination between the different layers.

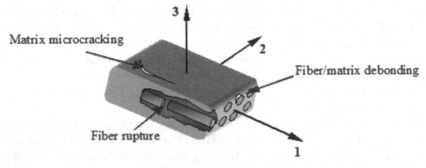

a)   Global composite ply damage

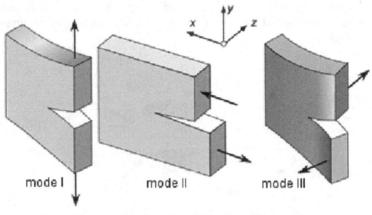

b)   Crack modes for crack initiation and growth

Fig. 2. Global composite ply damages [1] and basic forms of crack loading (applied to delamination growth) according to IRWIN [9]

3   FRP Simulation with VPS

Since FRP material is highly versatile, the numerical modelling and simulation possibilities and techniques are coherently complex. Bussadori et al. [7] gives an overview on currently available numerical models – especially for FRP simulation – which are described to some extent in the following paragraphs.

The VPS solver [1] can predict several failure models, element types and also crushing. In this context, it is profitable to use unidirectional (UD) plies due to the variability to build up many different models with less PLY cards. This is based on the need for numerous values and parameters for each PLY card, of which some are difficult to determine precisely. This implies the extensive time and cost needed to characterize a FRP for numerical crash simulation. The standard material model MAT131 is unable to calculate delamination but with the use of stacked shell models – i.e. one shell layer for each FRP ply – in combination with TIEDs as 1D interface elements to model delamination, this deficit can be fixed at computational costs expense. More detailed information regarding the numerical modelling and simulation of FRP within VPS can be found in [1, 8].

To provide better comparability, all force values – taken from experimental testing and numerical simulation – are normalized to the maximum value (peak force) of the corresponding force-displacement-curve and represented without any filters. Some work has been published to address these aforementioned deficits (e.g. [3]) but there is currently no suitable modelling technique available to handle the crushing failure mechanism in VPS. The work presented here provides a contribution to this challenging topic and offers a new modelling technique which is presented in more detail in the next section.

4   New Modelling Approach for FRP Crushing in VPS

In the context of this work, the explicit finite element solver VPS [1] was applied to finite element (FE) simulations of tri-axial braid composite structures in focus of applying the new modelling approach to automotive related specimens.

The basis for the numerical model are shell elements which are arranged in stacked order, i.e. the laminate is subdivided into several layers while each of those layers is modelled by one separate shell element surface. An enhanced Bevel-Trigger was used to initiate the crushing process. This was done on the one hand by cutting off the layers at different heights and on the other hand including a debris wedge which is switched off short time after simulation begin, as shown in Fig. 6. Additionally, TIED-elements are used to join the shell layers with each other and a bi-linear failure model has been applied for delamination modelling within the structure. This is done by a cohesive interface in combination with material card 303. The material itself is modeled as a stacked shell type material using material card 131. The applied material cards/types of the model as well as their main role and properties are as follows [1]:

- material 131: LADEVEZE model and definition of ply stacking (built up of laminate) in combination with SHELL elements
- ply type 1: unidirectional (UD)-ply containing FRP parameters referenced by the stack definition of material 131
- PUCK's and YAMADA SUN criteria for modelling fiber and/or matrix failure
- material 303: modelling delamination with PICKET model in combination with 1D-TIED elements

The bottom of the specimen is fixed regarding all degrees of freedom (DOF). Analyzed output parameters are the section force which is defined in the tube structure to determine the total force of the model and the crushing length. The mesh size was kept constant to the earlier investigations [2, 11] with 1 mm in general for the finite elements in the final model (Fig. 3).

Since it is not yet possible to work with braided fibers within the VPS solver, a work-around using unidirectional plies with different angles for each fiber layer is devised. With this method, good results with respect to the phenomenology of crushing damage can be observed.

Fig. 3. Numerical model and details regarding the debris wedge

When the impactor touches the model, the initiation and propagation of crushing through the plies begins. This is modelled with help of PICKET model (material 303) and LADEVEZE model (material 131). Inner failure is implemented using the LADEVEZE model, PUCK's and YAMADA SUN criteria. Stacked shell layers and delamination between the plies are depicted in a cross section view in Fig. 9. For more details concerning the model and its theoretical background see [2, 7, 8, 10] (Fig. 4).

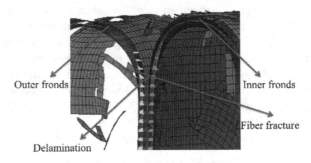

Fig. 4. Failure characteristic between layers

As mentioned earlier, the FRP material data (parameters) used for the numerical simulations of this work is obtained or improved respectively via a standard procedure of Volkswagen Group extended by additional steps regarding the new modelling approach presented here. As an example, Fig. 5 shows the results of a 1D compression test and a 4-point-bending test. The tests were carried out on the basis of DIN ISO EN 14126 and DIN ISO 14125 for each type (braid and axial yarn). For the compression test the specimen geometry was modified to compensate the influence of axial yarns.

An innovative modelling strategy has been developed to take special physical phenomena of tri-axial braids into account. This strategy focusses on the layer thickness and the corresponding Steiner-part (distance between symmetry plane and outer layer) on the one hand and on the fiber volume fraction (FVF) on the other hand.

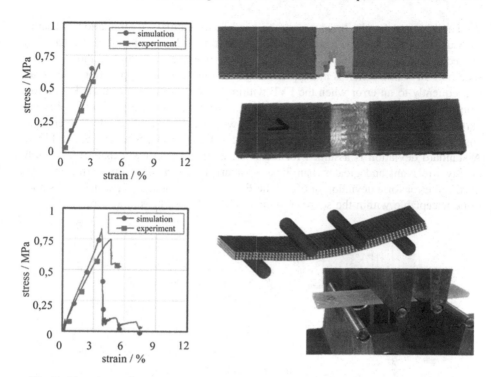

Fig. 5. Experimental and numerical results of 1D compression and 4-point-bending tests

The theoretical (numerical) layer thickness results from the specimen wall thickness of 3.2 mm divided by the number of layers which is 4 to 0.8 mm. Due to the fact, that within tri-axial braids the single layers intersect each other because of their structured surface, the "stiffness" layer thickness is 1.3 mm (see Fig. 6). Therefore, an engineering approach has been developed by adjusting the Steiner-part of the braid yarn for the stiffness calculation as mentioned before and remaining the part thickness at 3.2 mm, e.g. for global stability.

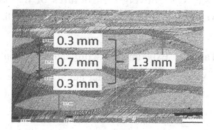

Fig. 6. Theoretical (numerical) layer thickness and "stiffness" layer thickness

The ply card for the braid yarn to be used in the numerical simulation is characterized with a fiber volume fraction (FVF) of about 50%. Due to the effort that has to be carried out to get those ply cards it is common practice to formulate those for one specific FVF value which is in a typical range for the requested application. This leads consequently to an error when the FVF within the test specimens or parts is differing from the FVF used to formulate the ply card. Unfortunately, the FVF within braided parts varies with the braid angle as it is shown in Fig. 7. To evaluate this error an engineering approach has been developed on the basis of Gaussian curve, see Fig. 7. A standard deviation is assumed for the FVF; a FVF value lower than 30% is technically irrelevant and greater than 70% is (nearly) impossible. Considering this, the maximal error for a deviation of 6% in the FVF value is about 12%, which is assumed to be acceptable within the scope of numerical FRP crushing simulation.

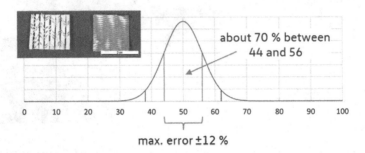

Fig. 7. FVF variation due to braid angle and engineering approach

5   Description of the Experimental Study

The experimental investigations are focused on FRP crash specimens with E-glass fibers inside a thermoset matrix. A braiding process was chosen to build up a tri-axial-braided specimen with a nearly rectangular cross section. *Crushing* initiation was done by a Bevel-Trigger.

The experimental testing was performed with a drop weight tower located at the Institute of Aircraft Design & Lightweight Structures of the TU Braunschweig, Germany. Each specimen was tested with parameters described in Table 1.

Table 1. Experimental setup and execution

Mass [m]	Velocity [v_{imp}]	Impact angle [α]	Observed data
0.2 t	4.4 m/s, 7.2 m/s	0° ... 20°	F, t, l, v, a

The force was measured by using a 3D quartz sensing element type 9077C from Kistler Instruments with components in x-, y-, and z-direction. No crash filters were used. The applied impact mass was fixed to a slide and test results show clear measurement results and high reproducibility.

For experimental evaluation, the key values of crushing efficiency, the specific energy absorption (SEA) and the average force (F_{ave}) are gathered. SEA is calculated according to equation [1], with the density ρ, cross section A, crash length 1 and crushing force F [3, 5, 8].

$$SEA = \frac{1}{\rho Al} \int_0^{l_{max}} F \, dl \tag{1}$$

More detailed information regarding the performed experimental study can be found in [9].

6 Results of the Numerical Model Validated with the Experiments

As mentioned earlier, a large experimental study has been carried out to validate the simulation results of the new modelling approach. The numerical data was evaluated regarding the modelling quality which is described in detail below. Besides an enhanced numerical stability is observed, indicated by a constant time step of 8.0×10^{-5} ms. The CPU time of the model is 2:41:35 h by using 64 CPUs ($l_{ele} = 1$ mm; $n_{ele} = 201.306$; $t_{crash} = 25$ ms).

The evolution steps in the context of the new modelling approach done so far and also those which are planned in the future are presented in Fig. 1. The *MSL-MCMC* approach, which is described in more detail in Hennemann et al. [8], uses experimental data and continuous comparison of those with the numerical results to improve the model parameters, especially material data. Initially, more than 15 parameters were considered, but after a sensitivity study, those could be reduced to the influential ones [8]. Within the further developed approach (*MSL-Strategy*) presented in this paper, the numerical results are compared only once with the experimental data to improve the model. The parameters which are changed after that, are the material parameters due to a differing FVF and the Steiner-part of the braid yarn for the stiffness calculation (modelling strategy, see Sect. 5). Obviously, if the FVF is not differing, the fitting is reduced to the Steiner-part issue. The comparison between the first numerical calculation and the experimental data can even be skipped, when both aforementioned issues are directly implemented into the numerical model just at the model built up. Looking forward to the final stage of the modelling approach (*M-L-A*; see Fig. 1), no fitting of the model has to be performed and therefore no comparison between numerical and experimental results has to be carried out.

For example, Fig. 8 shows the results for a drop weight tower test with an impactor angle of 0° which is perpendicular to the cross section of the specimen. The test specimen has a [± 68°, 0°, ± 68°]₄ laminate stack with a FVF of 53%. The energy absorption is 5.6 kJ. The experimental force-displacement-curve is reproduced very precise by the numerical model regarding the shape and the crushing force level. The peak value shows some deviations. Nevertheless, due to the local appearance of this value combined with a very sharp curve in this area, this deviation is assumed acceptable.

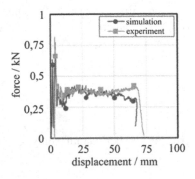 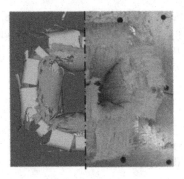

Fig. 8. Results for impactor angle of 0°

The failure mode which is provided by the numerical model shows the identical properties as the test specimen does. The delamination modelled by the cohesive elements occurs only in the upper part of the specimen (turquoise bars) and leads to a splaying mode that divides the single layers of the stacked-shell model. The laminate behavior within the corners, which is significantly different to the behavior on the longitudinal sides, is predicted accurately.

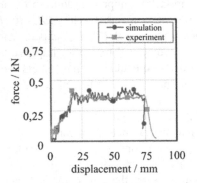 

Fig. 9. Results for impactor angle of 10°

Figure 9 shows the results for a drop weight tower test with an impactor angle of 10° referring to the cross section of the specimen. The test specimen has the same properties and energy absorption as mentioned earlier. The experimental force-displacement-curve is again reproduced very precisely by the numerical model regarding the shape and the crushing force level. The calculated peak value corresponds entirely with the experiment in this configuration.

The qualitative failure mode is provided by the numerical model with identical properties as the test specimen shows. This is remarkable, because most of the currently available approaches and models have large difficulties with these complex mechanical loadings especially when it comes to oblique impacts. The splaying mode due to delamination divides the single layers of the stacked-shell model even on the upper and lower longitudinal sides where the whole laminate is bending to the outer and inner side respectively. Within the corners the laminate behavior is quite accurately represented. This is the case for all corners irrespective of the splaying layers and the direction of bending.

7 Conclusion and Future Work

The presented work offers a new modelling technique for the appropriate numerical simulation of FRP *crushing* within finite element method (FEM) codes, such as VPS [1], since these codes currently have deficits in this context. The new approach provides a vital contribution to this challenging topic with the goal of improving numerical simulations for future vehicles. The accurate predictions of FRPs have significant advantages for the lightweight design of cars. Both, the numerical models and experiments were compared and evaluated. The new approach is able to model the *crushing* mechanism phenomenologically correct and further also predict the quantitative values and curve character accurately. Encouraged by the presented satisfactory results, the authors are extending this new modelling and simulation technique to further tasks set up on the already investigated ones, e.g. reduced number of stacked shells.

Parallel to the further development towards the final stage of the new modelling approach (*M-L-A*; see Fig. 1), a numerical study shall be carried out to investigate and proof the predictability of the modelling approach within its current stage (*MSL-Strategy*). This will also be done to have a fast work around and an alternative to the *M-L-A* approach.

References

1. Engineering Systems International (ESI). Virtual Performance Solution (VPS) (2014)
2. Honnakkalavar, H.: Crash investigations of composite structures using an explicit FEM-Solver and identification of failure parameters by stochastic method. Master Thesis. Wolfsburg. Restricted until 2023 (2018)
3. Feindler, N.: Charakterisierungs- und Simulationsmethodik zum Versagensverhalten energieabsorbierender Faserverbundstrukturen, München (2014)
4. Greve, L., Pickett, A., Payen, F.: Experimental testing and phenomenological modelling of the fragmentation process of braided carbon/epoxy composite tubes under axial and oblique impact. Compos. Part B: Eng. 39(7–8), 1221–1232 (2008)
5. Joosten, M., Dutton, S., Kelly, D., Thomson, R.: Experimental and numerical investigation of the crushing response of an open section composite energy absorbing element. Compos. Struct. 93(2), 682–689 (2011)
6. McGregor, C., Vaziri, R., Xiao, X.: Finite element modelling of the progressive crushing of braided composite tubes under axial impact. Int. J. Impact Eng 37(6), 662–672 (2010)
7. Bussadori, B., Schuffenhauer, K., Scattina, A.: Modelling of CFRP crushing structures in explicit crash analysis. Compos. Part B: Eng. 60, 725–735 (2014)

8. Hennemann, S., Hohm, V., Horst, P., Honnakkalavar, H.: Numerical treatment of fiber reinforced plastic crushing mechanism in virtual performance solution: possibilities and limitations. Proc. IMechE Part L: J Mater. Des. Appl. (2019). https://doi.org/10.1177/1464420719852968
9. Brocks, W.: Plasticity and fracture mechanics. Manuscript. Institute for Mechanic. TU Berlin, Berlin (2012)
10. Twardy, L.: Numerical analysis of failure mechanisms of fiber reinforced plastics and metal structures under crash load. Master Thesis. Wolfsburg. Restricted until 2024 (2019)
11. Zimmermann, P.: Dynamic and material interactions between fiber reinforced plastics and metal structures under crash load. Master Thesis. Wolfsburg. Restricted until 2023 (2018)

Reinforcement the Seismic Interaction of Soil-Damaged Piles-Bridge by Using Micropiles

Mohanad Talal Alfach[✉]

Faculty of Science and Engineering, School of Architecture and Built Environment, University of Wolverhampton, Wolverhampton, UK
mohanad.alfach@wlv.ac.uk

Abstract. This paper presents Three-dimensional numerical modeling of Soil-damaged Piles-Bridge interaction under seismic loading. This study focuses on the effect of developing plastic hinges in piles foundation on the seismic behavior of the system. In particular, this study is interested in evaluating the proposed approach for strengthening the system of soil-damaged piles-bridge. The proposed approach is based on using the micropiles whose significantly promoting the flexibility and the ductility of the system. This study was carried out using a Three-dimensional finite differences modeling program (FLAC 3D). The results confirmed the considerable effect of the developing of concrete plasticity in the pile's foundation, which reflects in changing the distribution of internal forces between the piles, also, they show the efficiency of using the micropiles as a reinforcement system. The detailed analysis of the micropiles parameters shows a slight effect of pile-micropile spacing, while the use of inclined micropiles leads to attenuation of the internal forces induced in the piles and the micropiles themselves.

Keywords: Interaction · Piles · Concrete · Seismic · Plasticity · Three-dimensional · Micropiles

1 Introduction

Piles foundations one of the most practical options used to enhance the structure's stability, but under severe seismic loadings they might subject to efforts exceed their allowable limit of seismic resistance. These efforts are particularly dangerous when the piles are implanted in nonlinear soils. The post-seismic analyses have revealed the fundamental role of soil-foundation-superstructure interaction on the seismic behavior of the piles and the structure (Kagawa and Kraft 1980; Mizuno et al. 1984; Boulanger et al. 1999; Miura 2002, …). Under a strong seismic loading, the nonlinearities of the soil and the structure can play a decisive role, consequently the plastic hinges are probably developed in the piles. In the literature, there are several models of concrete behavior, particularly Elastic-Perfectly Plastic model, other damage models may take into account the reduction in elastic rigidity and development of irreversible deformations. In the seismic design, the

© Springer Nature Switzerland AG 2020
N. Gunkelmann and M. Baum (Eds.): SimScience 2019, CCIS 1199, pp. 179–195, 2020.
https://doi.org/10.1007/978-3-030-45718-1_12

development of plastic hinges is allowed in specific locations (in heads as an example). In fact, these plastic hinges will absorb the oscillation induced by the seismic loading and thereby limit the resultants stresses. In this study, we analyse the influence of these plastic hinges on the seismic response of the system. The plasticity of the concrete piles is governed by a plastic moment M_p which is the maximum allowable bending moment of the pile. Cakir and Mohammed (2013) have studied the seismic retrofitting of historical masonry bridges using micropiles and have examined the seismic performance of this technique. Likewise, Momenzadeh et al. (2013) have confirmed the micropiles seismic retrofit efficiency in poor soil conditions with high structural load demands of the San Francisco Bay area Bridge. Furthermore, Doshi et al. (2017) have analyzed the seismic behavior of soil-piles-micropiles-bridge and confirmed the beneficial effect of the added micropiles in reducing the settlement, bending moment and shear force. Ousta and Shahrour (2001) have studied the seismic behavior of micropiles in saturated soils. Also, Sadek and Shahrour (2003) analysis showed that the inclination of micropiles results in an improvement of the lateral stiffness, the bending moment and the axial force of the piles. Similarly, Alsaleh and Shahrour (2009) have confirmed that the non-linearities of the soil and micropile-soil interface both have a significant effect on the seismic response of the micropiles group as well as that of the structure. The research conducted in this study provides a thorough analysis of the soil-pile-bridge under seismic loading. Particular attention was given to the influence of nonlinearity of concrete piles and the behavior of soil-pile-bridge reinforced by adding a group of micropiles. The study is carried out using Three-dimensional modelling code (Flac 3D).

2   Soil-Pile Structure System and Numerical Model

The model consists of an implanted group of piles in soil. The modeling of the behavior of such a system under seismic loading requires specific methods to take into consideration the interaction between those different components, namely the soil-piles, pile-pile, piles-cap interaction and all piles-cap-soil with the structure. The boundaries of the model should be put sufficiently away from the structure to minimize the effect of wave reflection which leads to dense mesh. To overcome this difficulty, we use specific borders that prevent them from reflecting on the model. FLAC 3D is used in this study, this code uses the Lagrangian representation of movement. It is based on the explicit finite difference method to solve the equations of dynamic equilibrium.

2.1   Reference Example: Elastic

The model consists of a bridge supported by a group of (6) floating reinforced concrete piles with the length of (Lp = 10.5 m). One layer of homogenous soil with a depth of (15 m) support the implanted piles, those piles are rigidly connected in a (e_c = 1 m) thick of the reinforced concrete cap as shown in Fig. 1. Tables 1 and 2 present the properties of the superstructure and the piles. Regarding the complexity of the problem of Seismic Soil-Structure-Interaction (SSSI), a number of measures were used:

- To avoid the pile–pile interaction, the spacing inter-piles were taken as (S = 3.75Dp = 3 m).

- To avoid the effect of soil-cap interaction, the cap was put in 0.5 m over the soil surface.
- The absorbent boundaries of the model have been used to minimize the effect of seismic waves reflection.
- A variable density mesh was employed to minimize the analysis cost, so we have densified the mesh at the center of the model where we expect important seismic soil-structure effects.

In this reference example, the behavior of soil-pile-structure is assumed to be elastic with Rayleigh damping, the factor of damping used is (5%) for the soil and (2%) for the structure. The fundamental frequency of soil is (0.67 Hz). The superstructure is modelled by a pillar supports a lumped mass in its head [m_{st} = 350 Tons]. The flexural modal rigidity of the superstructure is k_{st} = 86,840 kN/m and its frequency (assumed fixed at the base) are equal to F_{st} = 2.5 Hz. Those values were calculated by the following formulations:

$$k_{st} = \frac{3(E_{st} . I_{st})}{H_{st}^3}$$

$$F_{st} = 1/(2\pi)\sqrt{K_{st}/m_{st}}$$

The frequency of the superstructure taking in consideration the Soil-Structure-Interaction (SSI) is $f_{st,flex}$ = 1.1 Hz (including SSI).

ρ, E and v are the density, young modulus and the coefficient of poisson. ζ: is the factor of damping. Dp: is the pile diameter. E*A and E*I: are the axial and bending stiffness. The used mesh shown in Fig. 1(b) includes (3856) zones of (8) nodes and (138) three-dimensional beams of 2 nodes. The mesh was refined around the piles and near the superstructure where inertial forces induce high stresses.

Table 1. The elastic property of the soil and piles materials

Material	Diameter (m)	Mass density ρ (Kg/m³)	Young modulus E (MPa)	Poisson ratio v	Damping ratio ξ (%)	Height (m)
Pile	0.8	2500	20000	0.3	2	10
Soil		1700	8	0.45	5	15

Table 2. The elastic property of the superstructure

ρ_{st} (Kg/m³)	E_{st} (MPa)	v_{st}	ξ_{st} (%)	m_{st} (t)
2500	8000	0.3	2	350

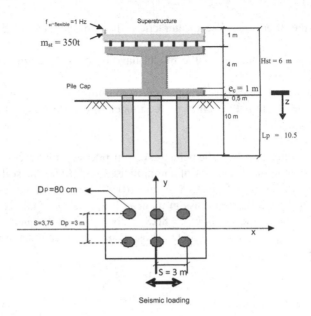

a) **System Geometry**

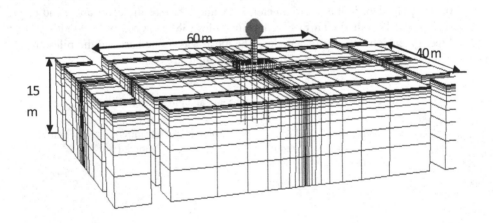

b) **3D numerical mesh with absorbing boundaries (138 beam elements and 6978 nodes)**

Fig. 1. Problem under consideration

2.2 Seismic Loading (Real)

The seismic loading chosen in this research is the one recorded in Kocaeli at Turkey on 17/08/1999 (Station AMBARLI; KOERI source). The seismic loading is applied as a speed at the base of the soil as shown in Fig. 2. The maximum amplitude of this loading is (40 cm/s) (acceleration maximum = 0.247 g). As the seismic numerical analyses are very costly in terms of time and hardware, so it was very helpful to carry out the numerical analysis for a shorter seismic record, hence all following analyses have carried out for Kocaeli seismic record of 1999, but for the duration of (time = 10 s), this choice was taken after an accurate check process to ensure that the effect of this seismic loading for the duration (time = 10 s) perfectly equalizes the effect of the whole seismic record duration (time = 30.08 s).

The spectrum of Fourier corresponding to the used seismic loading illustrated in Fig. 2. We note that the frequencies involved are less than (3) Hz with a maximum peak for (F = 0.9 Hz) which is between the fundamental frequency of the soil (F1 = 0.67 Hz) and the frequency of structure ($f_{st,flex}$ = 1.1 Hz), Hence, the choice of this loading in our analysis. Also, note that a first peak is observed for frequency (F = 0.6 Hz) which is very close to the fundamental frequency of soil.

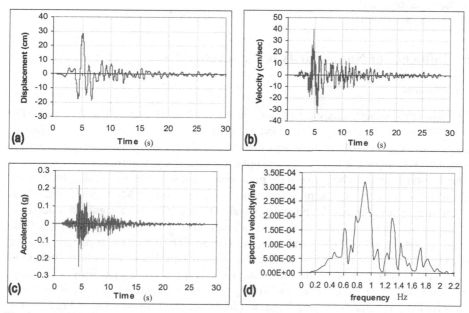

Fig. 2. Kocaeli earthquake record (1999). (a) Displacement, (b) Velocity, (c) Acceleration, (d) Fourier spectra of velocity component.

Table 3 shows the efforts induced in the piles under Kocaeli earthquake loading.

In order to compare the obtained results, the induced efforts are normalized to inertial forces of the superstructure as follows:

$$T^* = \frac{T}{T_{st}}$$

Table 3. Reference example: response of a group of (2*3) piles under Turkey loading (1999).

Seismic loading	a_{st} (m/s^2)	a_{Cap} (m/s^2)	Internal forces				Normalized forces	
			Central piles		Corner piles		Corner piles	
			T_{max} (kN)	M_{max} (kN.m)	T_{max} (kN)	M_{max} (kN.m)	T^*_{max}	M^*_{max}
Turkey	11.28	8.385	675.8	954.4	1016.1	1099	0.196	0.05

$$M^* = \frac{M}{m_{st} a_{st} H_{st}}$$

$$T_{st} = m_{st} * a_{st}$$

Where:

m_{st}: the mass of the superstructure.

T_{st} and α_{st} denote the shear force induced at the superstructure and the acceleration of the superstructure.

H_{st}: Superstructure height

3 Influence of Nonlinearity of Concrete Piles

3.1 Results

The numerical simulations were carried out for the case of frictional soil (C = 2 kPa, $\varphi = 30°$, $\psi = 20°$) under seismic loading recorded in Turkey (1999) with maximum amplitude ($V_{max} = 0.4$ m/s) the soil behavior is described by an elastic-plastic law without hardening according to the elasto-plasticity standard model of [Mohr–Coulomb]. The results of the linear behavior of the piles were compared with the results obtained for elasto-plastic behavior of the piles and plastic bending moment Mp = 500 kN.m which is the maximum moment the piles can stand. The results show a significant decrease in the internal forces induced in the piles with the nonlinear behavior of the piles. This result is figured out in Fig. 3 and Table 4. The acceleration of the superstructure for elastic pile behavior is (23.5%) higher than that obtained for elastoplastic behavior, While the comparison of elastic and elastoplastic responses (Table 4) reveals a reverse trend for the acceleration at the cap. The profile of the bending moment shows that the plasticity has attained over a large part of the pile and was not located only in the pile head (plastic hinge), this result indicates that the collapse of the structure is very probable with the elastoplastic behavior of the piles. On the other hand, we note that only in case of nonlinear behavior of piles, we obtain a reduction of maximum normal forces by about (24.6%) accompanying with a reduction of about (34%) for maximum shear forces in corner piles. Also, there is an attenuation in the spectral amplitude of the structure velocity as shown in Fig. 4, this is due to the damping induced by the plasticity of the piles. It is important to note that the presence of plasticity in the piles affects the distribution of the maximum shear forces between the central and corner piles.

Table 4. Influence of the nonlinear behavior of concrete pile on dynamic forces in piles. (frictional soil, Earthquake of Turkey, $V_g = 40$ cm/s)

Model (Concrete)	a_{st} (m/s²)	a_{Cap} (m/s²)	Dynamic forces						Normalized forces	
			Central piles			Corner piles			Corner piles	
			Nmax (kN)	Tmax (kN)	Mmax (kN.m)	Nmax (kN)	Tmax (kN)	Mmax (kN.m)	T^*_{max}	M^*_{max}
Elastic	9.567	6.592	38.2	511.7	897.9	1840.2	917.3	1140	0.211	0.062
Plastic Mp = 500 kN.m	7.744	8.327	38.8	450.8	500	1386.2	604.1	500	0.154	0.033

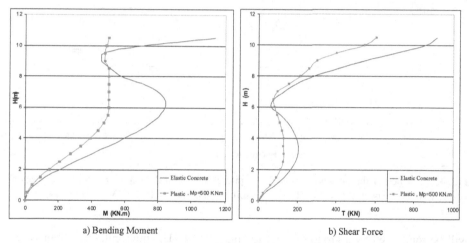

a) Bending Moment   b) Shear Force

Fig. 3. Influence of the non-linear behavior of concrete pile on dynamic forces in the corner piles. (Frictional soil, Earthquake of Turkey, $V_g = 40$ cm/s)

4 Piles Interaction Reinforcement

The site observations of (Lizzi and Carnevale 1981; Pearlman et al. 1993; Mason 1993; Herbst 1994), as well as recent research, have demonstrated that the micropiles system constitutes a reliable tool as reinforcement technique for existing structures. The facility of their installation, especially in difficult access locations which represent their main asset. Their use in the seismic sites provides great benefits because this system of foundations is characterized by good flexibility and ductility which are properties very appreciated for structures exposed to seismic risks. In the field of numerical modeling, the main research about the seismic behavior of micropiles was focused on their use as a foundation of new structures (Sadek 2003; Alsaleh 2009). In this part, we examine the response of the soil-pile-structure system reinforced by a group of micropiles. The used model is an identical system that previously studied with a foundation of (6) piles, that

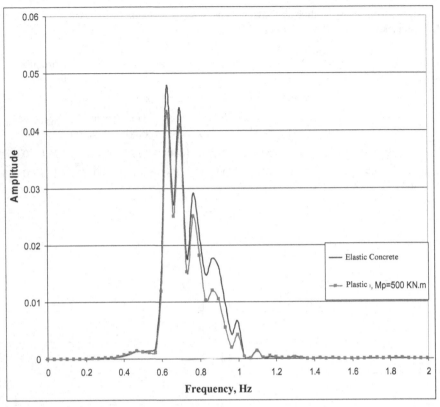

Fig. 4. Influence of the nonlinear behavior of pile concrete on the superstructure head spectral velocity (Fourier spectra diagram) (Frictional soil, Earthquake of Turkey, $V_g = 40$ cm/s)

will be reinforced by a group of (4) micropiles (2*2). The micropiles diameter is $D_m = 0.25$ m (Fig. 5). The implementation of the used reinforcement solution takes place by enlarging the existing cap in which the micropiles will be built in order to rigidify the foundation system. The interface between the piles or micropiles and the soil is supposed elastic. The calculations were carried out with assuming the value of piles plastic moment $Mp = 500$ kN.m. The applied loading is the record of Turkey (1999), but with a maximum amplitude ($V_{max} = 0.6$ m/sec). This amplitude was chosen in order to induce the development of plasticity over a large part of the piles, which justifies the reinforcement. The soil characteristics are identical to that used in the previous sections (Frictional soil: C = 2 kPa, $\varphi = 30°$, $\psi = 20°$). In order to limit the number of reinforcement elements (4 elements) and perform a qualitative analysis, the behavior of the micropiles assumed to be linear elastic, even if the induced forces exceed their bearing capacity. For a real study, the number of micropiles must be optimized according to applied seismic loading.

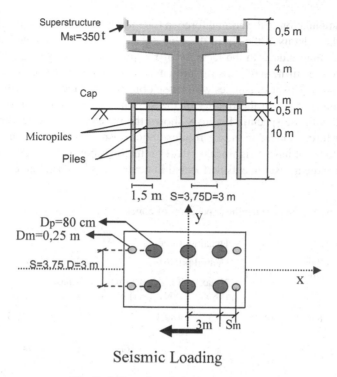

Seismic Loading

Fig. 5. Micropile reinforcement scheme

4.1 Effect of Pile-Micropile Spacing

In order to analyze the influence of pile-micropile spacing on the seismic response of the system, the numerical simulations were carried out for several values of the pile-micropile spacing (Sm = 1.5, 2 and 3 m). Tables 5, 6 and Fig. 6 give the results of the comparison between system response before and after reinforcement for several pile-micropile spacing's. Firstly, it is noted that the reinforcement by micropiles can constitute an effective reinforcement solution. In fact, there is a strong reduction of the internal forces in the piles after the reinforcement. Figure 6 compares the bending moment envelopes in the piles with and without reinforcement. After the reinforcement, we note that the plasticized area of the pile is reduced to a point located at the head of the pile, which can be accepted by seismic codes and does not jeopardize the structure. Furthermore, it can be seen that the maximum bending moment decreases with the increase of the pile-micropile spacing. A similar tendency is observed for the envelope of shear force in the pile. The pile-micropile spacing incites a reduction of the normal force of about (40%) in the micropile head. The influence of pile-micropile spacing is not very significant despite a small decrease of the internal forces with the spacing increase. In terms of the normal forces induced in the piles, the reinforcement was very beneficial in reducing the forces taken by the piles. In fact, a 75% attenuation of the maximum normal force in the piles after the reinforcement is obtained. However, the influence of the pile-micropile spacing is not important on the normal force in the piles, but it incites a significant attenuation of the normal force in the micropiles. By examining the spectral response of the velocity at the top of the superstructure (Fig. 7), we note a decrease in the maximum amplitude with the increase of the pile-micropile spacing. This reflects an increase in the stiffness of the structure and explains the reduction of the dynamic forces in the piles. This result agrees with those obtained by Sadek and Shahrour (2003) concerning the increase in the rigidity of the system and the reduction of the lateral amplification with the spacing. Figure 8 shows the envelope of the bending moment and the shear force induced in the micropiles under seismic loading. It can be seen that the spacing does not have a significant effect on the distribution of these forces. These results confirm the measurements performed in centrifuges by Nishitani et al. (2001).

Table 5. Influence of the nonlinear behavior of concrete pile on dynamic forces in piles

Model (Concrete) $Mp = 500$ (kN.m)	a_{st} mass (m/s^2)	a_{cap} Cap (m/s^2)	Dynamic forces					
			Central piles			Corner piles		
			Nmax (kN)	Tmax (kN)	Mmax (kN.m)	Nmax (kN)	Tmax (kN)	Mmax (kN.m)
Before the reinforcement	8.712	14.25	36.3	489.1	500	1553.2	691.5	500
After S = 1.5 m	8.432	8.4	8.6	289.1	500	376	359.6	500
After S = 2 m	8.358	8.759	0.5	281.9	500	422.6	339.7	500
After S = 3 m	7.608	8.131	0.4	247.2	500	331	291	500

Table 6. Influence of the pile-micropile spacing on the dynamic forces in the reinforcing micropiles.

Model	Dynamic forces (micropile)		
	Nmax (kN)	Tmax (kN)	Mmax (kN.m)
S = 1.5 m	984	602.7	1090
S = 2 m	841.3	596.6	1080
S = 3 m	594	491.4	926.8

4.2 Effect of Micropile Connection

In this section, we analyze the effect of the micropile-cap connection on the seismic response of the system. Two types of connections are studied: fixed and articulation. The analysis is carried out for pile-micropile spacing (S = 2 m), Figs. 9, 10 and Tables 7, 8 present the results obtained for the two studied cases. By checking the induced forces in the piles, we note that the reinforcement by fixed micropiles reveals better efficiency

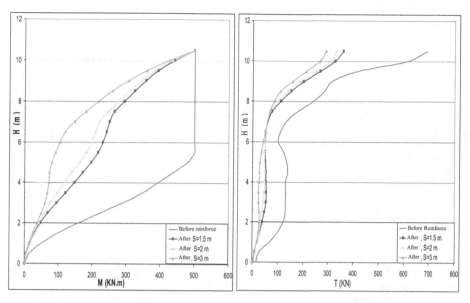

Fig. 6. Influence of the pile-micropile spacing on dynamic forces in the corner piles.

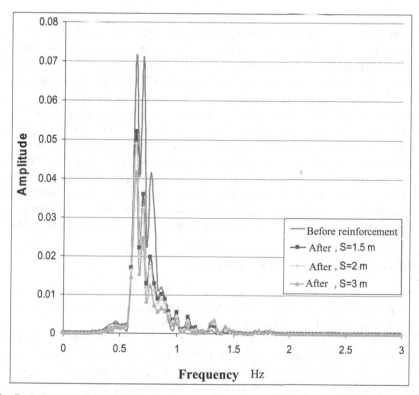

Fig. 7. Influence of the pile-micropile spacing on the superstructure head spectral velocity.

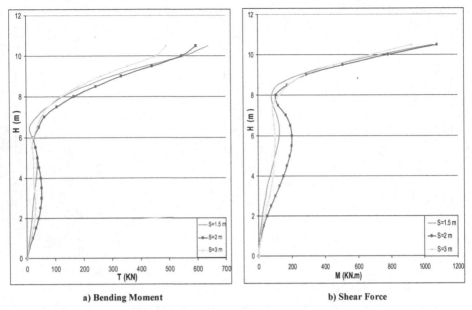

a) Bending Moment   b) Shear Force

Fig. 8. Influence of the pile-micropile spacing on the dynamic forces in the reinforcing micropiles.

in comparison with the articulated micropiles. For the articulated micropiles, the plastic moment has attained the upper quarter of the pile. Furthermore, the maximum shear force of the piles obtained in the case of articulated micropiles higher by (25–35%) than the obtained in the case of reinforcement by fixed micropiles. This result accompanied by the increase of the cap acceleration, which reflects lower rigidity for the system in the case of reinforcement by articulated micropiles. The normal force shows comparable maximum values in both cases (Fixed, articulated). Concerning the internal forces developed in the micropiles, the presence of the articulation permits to relieve the induced internal forces in the pile head. In that case, the maximum moment obtained at a depth of (2 m) of the micropile is considerably inferior to that developed at the top of the fixed micropiles M = 1080 kN.m, this result is consistent with the results found by Sadek and Shahrour (2003). Similarly, the profile of the shear force shows a decrease in the case of articulated micropiles, which is not in favor of the internal forces induced in the existing piles.

4.3   Effect of the Inclination of the Micropiles

In order to analyze the effect of the micropiles inclination on the seismic response of the system of bridge-soil-bridge, the calculations were made for pile-micropile spacing (S = 2 m). The soil-pile and soil-micropile connections are assumed to be perfectly rigid. Figures 11, 12 and Tables 9, 10 present the results of the numerical simulations carried out for two inclinations of the micropiles ($\alpha = 0°$, $\alpha = 15°$) for the case of frictional soil (C = 2 kPa, $\varphi = 30°$, $\psi = 20°$). In the case of inclined micropiles, we note a considerable decrease in the shear force accompanied by attenuation of the lateral acceleration at the superstructure and the cap. This decrease in the shear force is of

Table 7. Influence of the micropile-cap connection on the dynamic forces in the piles. (frictional soil ($\varphi = 30°$, C = 2 kPa, $\psi = 20°$), Earthquake of Turkey, $V_g = 60$ cm/s)

Model (Concrete) Mp = 500 kn.m	a_{st} mass (m/s^2)	a_{cap} Cap (m/s^2)	Dynamic forces					
			Central piles			Corner piles		
			Nmax (kN)	Tmax (kN)	Mmax (kN.m)	Nmax (kN)	Tmax (kN)	Mmax (kN.m)
Fixed micropiles	8.358	8.759	0.5	281.9	500	422.6	339.7	500
Articulated micropiles	8.179	14.28	18.7	381.2	500	406.1	499.4	500

Table 8. Influence of the micropile-cap connection on the dynamic forces in the micropiles. (frictional soil ($\varphi = 30°$, C = 2 kPa, $\psi = 20°$), Earthquake of Turkey, $V_g = 60$ cm/s)

Model	Dynamic forces		
	Nmax (kN)	Tmax (kN)	Mmax (kN.m)
Fixed	841.3	596.6	1080
Articulated	655.8	359	529.7

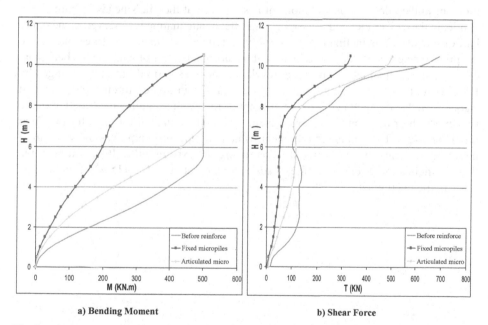

a) Bending Moment   b) Shear Force

Fig. 9. Influence of the micropile-cap connection on the dynamic forces in the corner piles. (frictional soil ($\varphi = 30°$, C = 2 kPa, $\psi = 20°$), Earthquake of Turkey, $V_g = 60$ cm/s)

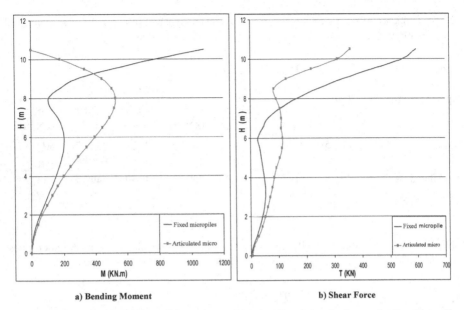

a) Bending Moment   b) Shear Force

Fig. 10. Influence of the micropile-cap connection on the dynamic forces in the micropiles. (frictional soil ($\varphi = 30°$, C = 2 kPa, $\psi = 20°$), Earthquake of Turkey, $V_g = 60$ cm/s)

the order of (30%) compared with the vertical reinforcement. Concerning the bending moment in the piles, the plastic moment has attained at the pile's heads in both cases, While, in the case of inclined micropiles, significant attenuation is observed in the upper half of the piles. The inclination is also very beneficial for the normal force induced in the piles, where a considerable reduction of the normal force is obtained. This beneficial effect of the inclination confirms the results of the tests carried out in centrifuges by Fukui et al. (2001). By examining the forces induced in the reinforcing micropiles, it can be seen that the inclination of the micropiles results in a significant reduction of the maximum shear force and bending moment. This reduction attains (53%) for the shear force and (42%) for the bending moment, which significantly improves the strength of the reinforcement elements without jeopardizing the existing piles. It should be noted that the inclination of the micropiles leads to a moderate increase (15%) of the normal force in the micropiles.

Table 9. Influence of the inclination of the micropiles on the dynamic forces in the piles. (frictional soil ($\varphi = 30°$, C = 2 kPa, $\psi = 20°$), Earthquake of Turkey, $V_g = 60$ cm/s)

α (°)	a_{st} mass (m/s²)	a_{cap} Cap (m/s²)	Dynamic forces					
			Central piles			Corner piles		
			Nmax (KN)	Tmax (KN)	Mmax (KN.m)	Nmax (KN)	Tmax (KN)	Mmax (KN.m)
0	8.358	8.759	0.5	281.9	500	422.6	329.7	500
15	6.473	6.897	0.3	206.5	500	126	231.9	500

Table 10. Influence of the inclination of the micropiles on the dynamic forces in the micropiles. (frictional soil ($\varphi = 30°$, C = 2 kPa, $\psi = 20°$), Earthquake of Turkey, $V_g = 60$ cm/s)

α (°)	Dynamic forces		
	Nmax (kN)	Tmax (kN)	Mmax (kN.m)
0	841.3	596.6	1080
15	978.5	276	624

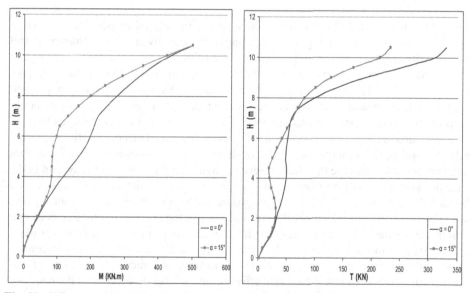

Fig. 11. Influence of the inclination of the micropiles on the dynamic forces in the corner piles. (frictional soil ($\varphi = 30°$, C = 2 kPa, $\psi = 20°$), Earthquake of Turkey, $V_g = 60$ cm/s)

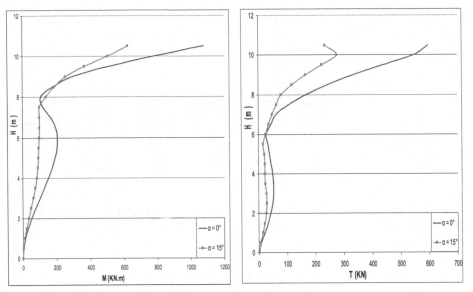

Fig. 12. Influence of the inclination of the micropiles on the dynamic forces in the micropile. (frictional soil ($\varphi = 30°$, C = 2 kPa, $\psi = 20°$), Earthquake of Turkey, $V_g = 60$ cm/s).

5 Conclusions

This study was devoted to global numerical modeling of soil-pile-bridge interaction problem under seismic loading. Attention particular was given to the influence of soil nonlinearity and development of plastic hinges in piles. The research within the domain of this study was conducted using a three-dimensional finite differences modeling program (FLAC 3D). The consideration of the non-linear behavior of concrete is important, especially if the bearing capacity of the concrete is likely to exceed. It permits better prediction of system failure under any load. The development of plastic hinges leads to an attenuation of the overall response of the system. It is found that the plasticizing of the piles change the report of the distribution of shear and normal forces between the central and corner piles. This type of behavior permits to predict the failure of the system in the case of the extension of the plasticity in the concrete piles, which is not the case with a linear behavior. On the other hand, it permits an analysis of the behavior of an existing structure requiring reinforcement. The results confirm the efficiency reinforcement system by micropiles for existing piles. The implantation of reinforcement elements plays a decisive role in the response of the system. Parametric study of the reinforcement system (Pile-micropile spacing, micropile connection, and micropile inclination) was carried out. The results reveal the slight effect of pile-micropile spacing on the response of the system. While the fixing of the micropiles in the cap allows a better attenuation of the forces in the piles and the structure. Similarly, the use of inclined micropiles as reinforcing elements is very beneficial to existing piles and micropiles themselves. These conclusions confirm the results of recent centrifuge tests on groups of piles reinforced by micropiles.

References

Alsaleh, H., Shahrour, I.: Influence of plasticity on the seismic soil-micropiles-structure interaction. Soil Dyn. Earthq. Eng. **29**(3), 574–578 (2009)

Boulanger, R.W., Curras, C.J., Kutter, B.L., Wilson, D.W., Abghari, A.A.: Seismic soil-pile-structure interaction experiments and analyses. J. Geotech. Geoenviron. Eng. **125**(9), 750–759 (1999)

Cakir, F., Mohammed, J.: Micropiles applications for seismic retrofitting of historical bridges. Int. J. Eng. Appl. Sci. **5**(2), 01–08 (2013)

Doshi, D., Desai, A., Solanki, C.: Bridge foundation restoration with retrofitting technique. Int. J. Civ. Eng. Technol. **8**(6), 856–866 (2017)

FLAC: Fast Lagrangian Analysis of Continua, vol. I. User's Manual, vol. II. Verification Problems and Example Applications, Second Edition (FLAC3D Version 3.0), Minneapolis, Minnesota 55401 USA (2005)

Herbst, T.F.: The GEWI-PILE, a micropile for retrofitting, seismic upgrading and difficult installation. In: International Conference on Design and Construction of Deep Foundations, Sponsored by the US Federal Highway Administration (FHWA), vol. 2, pp. 913–930 (1994)

Kagawa, T., Kraft Jr., L.M.: Seismic p-y response of flexible piles. J. Geotech. Eng. Div. ASCE **106**(GT8), 899–918 (1980)

Lizzi, F., Carnevale, G.: The static restoration of the leaning AL Hadba Minaret in Mosul (Iraq). In: Proceedings of the 3rd International Symposium of Babylon, Baghdad (1981)

Mason, J.A.: CALTRANS full scale lateral load test of a driven pile foundation in soft bay mud; Preliminary results. California Department of Transportation (CALTRANS), Division of Structure, California (1993). 31 pp.

Mizuno, H., Iiba, M., Kitagawa, Y.: Shaking table testing of seismic building-pile-two-layered-soil interaction. In: Proceedings of the 8th World Conference on Earthquake Engineering, San Francisco, vol. III, pp. 649–656 (1984)

Miura, F.: Some typical examples of damage to pile foundation by the 1995 Hyogoken Nambu earthquake. In: International Workshop on Micropiles - IWM 2002, Session VI, Detailed Seismic Design and Performance Issues, Venice, Italy, 29 May 2002–2 June 2002 (2002)

Momenzadeh, M., Nguyen, T., Lutz, P., Pokrywka, T., Risen, C.: Seismic retrofit of 92/280 I/C foundations by micropile groups in San Francisco Bay area, California. In: Seventh International Conference on Case Histories in Geotechnical Engineering, Chicago, 29 April 2013–4 May 2013 (2013). Paper No. 2.58

Nishitani, M., Umebar, T., Fukui, J: Development of seismic retrofitting technologies for existing foundations. In: Proceedings of the 5th international workshop on micropiles, Venice, Italy (2001)

Pearlman, S.L., Wolosick, J.R., Gronek, P.B.: Pin piles for seismic rehabilitation of bridges. In: Proceedings of the 10th International Bridge Conference, Pittsburgh (1993)

Ousta, R., Shahrour, I.: Three-dimensional analysis of the seismic behavior of micropiles used in the reinforcement of saturated soils. Int. J. Numer. Anal. Methods Geomech. **25**, 183–196 (2001)

Sadek, I., Shahrour, M.: Influence of boundary conditions on the seismic behaviour of micropiles foundations. In: International Workshop on soil-structure interaction - TC 4 Earthquake Geotechnical Engineering of the ISSMGE, Prague (2003)

Turkey, (Station AMBARLI; KOERI source) (1999). https://strongmotioncenter.org/vdc/scripts/plot.plx?stn=1911&evt=510

Author Index